Local and Urban Governance

Series Editor

Carlos Nunes Silva, Institute of Geography and Spatial Planning
University of Lisbon, Lisbon, Portugal

This series contains research studies with policy relevance in the field of sub-national territorial governance, at the micro, local and regional levels, as well as on its connections with national and supranational tiers. The series is multidisciplinary and brings together innovative research from different areas within the Social Sciences and Humanities. The series is open for theoretical, methodological and empirical ground breaking contributions. Books included in this series explore the new modes of territorial governance, new perspectives and new research methodologies. The aim is to present advances in Governance Studies to scholars and researchers in universities and research organizations, and to policy makers worldwide. The series includes monographs, edited volumes and textbooks. Book proposals and final manuscripts are peer-reviewed.

The areas covered in the series include but are not limited to the following subjects:

- Local and regional government
- Urban and metropolitan governance
- Multi-level territorial governance
- Post-colonial local governance
- Municipal merger reforms
- Inter-municipal cooperation
- Decentralized cooperation
- Governance of spatial planning
- Strategic spatial planning
- Citizen participation in local policies
- Local governance, spatial justice and the right to the city
- Local public services
- Local economic development policies
- Entrepreneurialism and municipal public enterprises
- Local government finance
- Local government and sustainable development
- Anthropocene and green local governance
- Climate change and local governance
- Smart local governance

The series is intended for geographers, planners, political scientists, sociologists, lawyers, historians, urban anthropologists and economists.

More information about this series at http://www.springer.com/series/16129

Riya Banerjee • Gopa Samanta

Negotiating Terrain in Local Governance

Freedom, Functioning and Barriers of Women Councillors in India

Springer

Riya Banerjee
School Education Department
Government of West Bengal
Burdwan, West Bengal, India

Gopa Samanta
Department of Geography
The University of Burdwan
Burdwan, West Bengal, India

ISSN 2524-5449 ISSN 2524-5457 (electronic)
Local and Urban Governance
ISBN 978-3-030-60662-6 ISBN 978-3-030-60663-3 (eBook)
https://doi.org/10.1007/978-3-030-60663-3

This Springer imprint is published by the registered company Springer Nature Switzerland AG
The registered company address is: Gewerbestrasse 11, 6330 Cham, Switzerland

This book is dedicated to our beloved parents
who never told us
'You are a woman and you cannot do this'

Preface

The women's movement in modern India has a long history, starting with the abolition of sati and women's access to education in the late nineteenth century. Since then, there have been numerous forms of women's movements, agitating for equal rights in different parts of the country. As a consequence, many things have changed in women's lives. However, despite the positive changes, there still exist many instances of gender injustice even today. According to the Gender Inequality Index 2017, India ranks 127th in gender equality out of 160 countries of the world. Social and cultural changes take time, and changing women's conditions always take more time than usual. That is why the women's movement believes that without organized pressure and effective training and support systems, no positive change can be brought into the lives of women in a big way.

Other kinds of social injustice in India, such as inequality based on caste, have been addressed by the constitution through reservation, but gender injustice has never been adequately addressed through reservation in any sphere, be it in education or in jobs. However, without those reservations, which are the basic pillars of human development, the participation of women in local governance has been ensured by the 73rd and 74th Constitutional Amendment Act. This Act became a landmark in the inclusion of women in local government.

At the initial stage, women had to struggle against two institutions: the family, and the political and government institutions. Popular media had a tremendous negative response to the practice of families trying to manipulate this Act by using women representatives as proxy for their husbands, fathers-in-law or sons, which undermined the meaning of reservation. Moreover, the intelligentsia and power elites were not yet convinced of the efficacy of reservations for women in governance. Women thus began their struggle on an uneven field and had to face a number of roadblocks while joining local-level governing bodies. The specific hurdles they faced included: (a) lack of familiarity with male politics, political models, and discourses; (b) restrictions on their mobility; (c) lack of education and dependency on male relatives or colleagues; (d) the function of bureaucracy; (e) problems in learning rules and regulations and procedures; and (f) lack of information and knowledge. However, these roadblocks could not push them back home. Women soon started developing their own ways and means to navigate through the traditionally male-dominated workspaces, such as municipal offices.

Since the beginning of reservation for women in local governance, many studies have sought to check the level of empowerment of women achieved through this reservation. One group of scholars (Pai 1998; Bari 2005; Beall 2010; Kodoth and Mishra 2011; Rai 2011; Kulkarni 2011; Palanthurai 2012) came up with the observation that nothing has changed for women, while the other group of researchers (Holzner and Wit 2003; Chattopadhyay and Duflo 2001/2003; Ruet and Lama-Rewal 2005; Omvedt 2005; Beaman et al. 2006; Ghosh 2009; Joseph 2011; Sathe et al. 2013, and others) claimed that although the changes were not very significant, some changes definitely have taken place. Most of these studies looked into either the case for women in rural local governance or for women in metropolitan or big city local governments.

This book differs from earlier studies on women in local governance in two ways. We have examined the cases of women in urban local governance, but in small cities, society is more orthodox than in metropolitan cities and consequently gender norms are more difficult to negotiate. The boundary between the private and public lives of women is much stronger and the two worlds are more segregated, thus leading to women's mobility in the public realm being much more limited. Secondly, the book does not intend to measure the level of empowerment of women after joining urban local governance. Rather, it emphasizes the journeys of these women, which we feel is more important than the end result measured through the indicators of empowerment. The book is thus a comprehensive attempt to analyze, discuss and assess how the socio-cultural spaces of women councillors change with their participation in urban governance and how they negotiate the new unknown terrain.

The entire book is presented in nine chapters in total. The first chapter introduces the book along with the conceptual background based on the existing literature, the central research question and the objectives, selection of study area for in-depth fieldwork, a brief introduction of the selected cities and the research process. The conceptual background is built around issues of women's participation in politics and governance, women's reservation,[1] rethinking proxies and women's agency, especially *Development as freedom* as conceived by Sen (2000) and *Having, loving, being: An alternative to the Swedish model of welfare* as developed by Allardt (1993). The research process illustrates the selection criteria and the methods of field research.

The second chapter sets the background of the field research by discussing the strategies and pathways of urban local governance in West Bengal. The state of West Bengal, selected for this field research, is located in the eastern part of India

[1] Because of the abysmally low representation of women in governance at both local and national levels, the issue of women's reservation in elected governments is a significant one in India. Even after seven decades of independence, only 14% of elected representatives in the Indian Parliament are women. To increase the participation of women in local governance (both in rural and urban areas), 33% reservation of seats for women was ensured in the election of local governments by the 73rd and 74th Constitutional Amendment Acts passed in the early 1990s. Following that reservation policy, the women are relatively better represented (one-third) in the local governance.

and is one of the progressive states for several different reasons. The state has a long history of decentralization even before the 74th Constitutional Amendment Act. The urban governance reforms led by the World Bank and other international institutions have also been followed quite early in the state. The chapter also analyzes how this decentralization is undermined by the political parties in power through a very strong centralized model of political structure controlling urban local governments.

The third chapter looks into the history of women's development and empowerment programmes in India with a special focus on women's reservation in urban local governance. The situation of women in local governance in India has also been discussed. Significant emphasis has also been given to the condition of women in local governance in West Bengal. To understand the current status of women in local governance in West Bengal, we have analyzed the ratio of observed and expected participation of women in local governance and their representative participation in different parties fighting elections in West Bengal. This chapter also examines the role of local history in emancipating women in local governance at present. Out of the four cities under study, Darjiling and Balurghat have a history of women participating in mass movements like the Gorkhaland and Tebhaga movements, respectively. This chapter thus illustrates that cities with a pre-existing history of women's agency and emancipation are better places for women to work as councillors and to actually wield this power effectively.

The fourth chapter represents a brief profile of the women and men councillors in the selected cities.[2] In this chapter, we basically try to understand: (a) who are the councillors (both men and women) and (b) what are their social, economic, cultural and political backgrounds. Without knowing the background of these councillors, it becomes extremely difficult to understand their freedom and functioning. The collection of basic information with the help of the questionnaire facilitated us to remain in contact and to develop a personal relation with them, which helped us in conducting a more detailed investigation of their position and emancipation through qualitative methods. Qualitative methods usually take longer, and it was because of our earlier efforts at building rapport and maintaining frequent contact with them that they were willing to spend their valuable time in interviews, long conversations, and in focus group discussions at their offices as well as at their residential premises. For the profile, we focussed on family background and personal information such as position, responsibilities, caste and age-wise distribution and nature of elected seats (i.e., reserved for women or not) especially for the women councillors. Further, the nature of representation (i.e., proxies or not) has been explained to gain a detailed overview of the situation of the local councillors with special reference to women.

---

[2] To maintain research ethics, names of the councillors (both men and women) are not used in the book. Instead, a number is used against the councillors. This number has been given on the basis of the concerned Ward from which they have been elected. For example, a woman councillor who is elected from ward number 18 is numbered as WC18. The same system is followed for men councillors as well.

The fifth chapter examines the changing socio-cultural spaces and quality of life of the women councillors linked to their participation in city governance. This has been explained with the help of 'Having, Loving, Being: An Alternative to the Swedish Model of Welfare' written by Erik Allardt in the book *The Quality of Life* edited by Martha C. Nussbaum and Amartya Sen. Time-use surveys have also been used to assess their changing time-use patterns due to their participation in urban governance. The analysis of time use shows that after joining as councillors, women face the double burden of household work and official responsibilities as councillor. A detailed analysis of the time-use survey with intersectional lens such as age, marriage and work helped us understand that women councillors cannot be treated as a fixed category of women, rather it is an amorphous and complex category. For example, unmarried women have less responsibility of domestic work and they can devote more time as councillor.

The sixth chapter evaluates the freedom and functioning behaviour of women councillors. Here, four types of freedoms are elaborated upon: political freedom, socio-economic freedom, transparency and protective security. Functions of women councillors are classified on the basis of their work, for instance, decision-making, developmental works, implementation of women's empowerment programmes, and so on. However, each category of freedom and functioning is analyzed city wise, which shows that the situation of women's emancipation varies widely over spaces, depending on the local socio-cultural conditionalities and the history of women's work and agency in other different activities.

The seventh chapter uses a gender lens to make comparisons between the work of women and men councillors, to understand the gendered nature of local representation in city management. This chapter also accommodates the analysis of the perception of male councillors on the capability of the women councillors to understand how patriarchy works to limit the capability of women.

The eighth chapter analyzes the citizens' perception and their assessment of the local representatives, both women and men councillors. The analysis helps us understand how long-standing patriarchal forces have affected the mind-set of the citizens, making them reluctant about the capacity of women to run the governance system. However, deeper analysis shows that citizens' perception of the capacity of women councillors also varies based on their socio-cultural and gender context. This chapter helps us understand the complexity hidden within the system of perceiving the capabilities of their local representatives.

The final chapter is the conclusion, where after discussing our findings on the capability and barriers faced by women councillors, and the gendered perception of citizens about their capacity, we try to show how intersectionality becomes an important factor in making the 'women councillor' a fluid category, rather than fixing them into one box opposite to the men councillors. The study shows how caste, education, family background, marital status, household size, past experience in politics and current affiliation to political party also become important factors that affect the capability of women councillors, especially in terms of freedom and functioning.

Burdwan, West Bengal, India Riya Banerjee

Burdwan, West Bengal, India Gopa Samanta

Acknowledgements

We wish to express our sincere thanks to the Chairman and the local government officials of the cities, who helped us to collect the data during the field work. We are extremely grateful to the women and men councillors of the cities and the citizens who spared their time for interviews and informal discussions. This book would not have been possible without their cooperation. We are also thankful to the UGC for Riya Banerjee's PhD Fellowship and The Ford Foundation and ICSSR for Gopa Samanta's sponsored research projects, under which we have carried out the fieldwork for this research. We are especially thankful to the women councillors who provoked us to write their stories, which finally pushed us to write this book.

We are indebted to the University of Burdwan for providing us with the infrastructure required to carry out our research. We are thankful to the academic community of the Department of Geography, University of Burdwan, and also of other institutions located in different other parts of India and abroad who have helped us by giving critical comments on our work whenever we have presented our findings on this particular issue in different seminars, conferences and workshops. However, despite the concerned efforts of all these persons, and their valuable comments, there may be some limitations and inadequacies, for which we take the sole responsibility.

Contents

About the Authors

Riya Banerjee was a Research Fellow in the Department of Geography, The University of Burdwan for five years ending in 2018. Presently, she is posted as a Sub-Inspector of Schools under the School Education Department, Government of West Bengal, India. Her research interests lie in the fields of gender studies, urban and environmental studies. She has carried out her research on women in local governance for more than five years and recently has been awarded a PhD in 2019. She has presented research papers in different national and international seminars and conferences. She has publications in national and international journals and in edited volumes as book chapters. Besides writing in English, she has also written in Bengali and in popular magazines to reach out to common people outside academia.

Gopa Samanta works as a Professor in the Department of Geography, University of Burdwan. Currently, she is also in additional charge as Director, UGC-HRDC, University of Burdwan. Her research interests cut across the fields of urban, gender and water studies. She has carried out research in Eastern India with interdisciplinary methodologies and collaborative research teams under financial assistance from different national and international institutions such as the ICSSR, Ford Foundation, World Bank, Australia India Institute and INR-France, IRD-France. As a geographer, she has passion for undertaking in-depth and field-based empirical research. She acted as 'Gender Chair' of Paris-Sorbonne University, France, in 2016. She was a visiting fellow at Paris Diderot University, France (May–June 2018); Australian National University, Australia (May 2018 and October–November–December 2004); and French Institute of Pondicherry (October 2005 and October 2006), India. Her co-authored book *Dancing with the River: People and Life on the Chars of South Asia* has been published from Yale University Press. She has contributed immensely through publishing journal articles and book chapters. She acts as peer reviewer of articles in academically acclaimed journals such as *Community Development Journal, Development in Practice, ACME: An International E-Journal for Critical Geographies, Economic and Political Weekly, Contemporary South*

Asia, *Local Environment*, and *Built Environment*. Her detailed profile on teaching and research is available at the Burdwan University website. She is dedicated to disseminating the knowledge of social science in general and geography in particular beyond the classroom and academic community. To reach a wider audience, she contributes to popular writing in provincial language (Bengali) through news articles and other popular magazine articles. She is passionate in mentoring PhD students coming from peripheral locations.

Abbreviations

AITC	All India Trinamool Congress
BJP	Bharatiya Janata Party
BOC	Board of Councillors
BPL	Below Poverty Level
CIC	Chairman-in-Council
CPI (M)	Communist Party of India (Marxist)
CPI	Communist Party India
CSWI	Committee on Status of Women, India
EWR	Elected Women Representatives
FMR	Female–Male Ratio
H.S.	Higher Secondary
GJM	Gorkha Janamukti Morcha
GJNM	Gorkha Janamukti Nari Morcha
GTA	Gorkha Territorial Administration
ICDS	Integrated Child Development Services
ILO	International Labour Organisation
INC	Indian National Congress
IPU	Inter-Parliamentary Union
KMDA	Kolkata Municipality Development Authority
MC/MCs	Men Councillor(s)
MGNREGS	Mahatma Gandhi National Rural Employment Guarantee Scheme
MLA	Member of Legislative Assembly
MNAs	Member of the National Assembly
MP	Member of Parliament
NDA	National Democratic Alliance
NGO	Non-Governmental Organization
OBC	Other Backward Classes
RSP	Revolutionary Socialist Party
SC	Scheduled Caste
SNA	System of National Accounts
SFI	Students' Federation of India

ST	Scheduled Tribe
SUDA	State Urban Development Agency
SSK	Sahbhagi Shikshan Kendra
UI	Urban Intensity
ULBs	Urban Local Bodies
UPA	United Progressive Alliance
UPE	Urban Poverty and Eradication
UNDP	United Nations Development Programme
WB	West Bengal
WC/WCs	Women Councillor(s)
WPR	Workforce Participation Rate

Chapter 1
Introduction

1.1 Introduction

Women represent half of the world's population; however, they are still being discriminated against in every sphere of life. Their social space is gradually changing with changes in specific gender relations. The term 'gender' itself has a long history and range of meaning. It was introduced from the Latin word 'la: genus' meaning 'type' and 'kind', and in Greek, 'gen' means 'to produce' (Raju and Lahiri-Dutt 2011: 1). In this context, it is important to mention that French philosopher, Simone de Beauvoir, in her book *The Second Sex* (1949) argued: 'one is not born, but rather becomes a woman' (Borde and Malovany-Chevallier 2009: 330). Therefore, 'gender' refers to how societies set the behavioural, social, and cultural rules for being a man or a woman. However, such behavioural norms may vary over space and time. For example, today many men and women do things which were not socially acceptable until very recently. Again the norms are not the same in every location. It not only varies between countries, but also in everyday spaces and interactions. Moreover, women have differential access to space and place; for instance, public and institutional spaces are overtly considered as masculine (Raju and Lahiri-Dutt 2011: 48).

Studying women is still problematic in developing countries, though the movement for gender identities had started much earlier. In different academic disciplines, the word 'man' has subsumed both men and women. Till the 1970s, gender remained invisible from many social science disciplines in India. The movement to overcome this invisibility had started earlier, with the claim of right to vote for women in 1848, in the Seneca Falls Convention in the USA. After that, the second wave of the movement started with the radicalization of women during the anti-Vietnam war and civil rights movements in the 1960s. This led to the inclusion of women in academic disciplines, but only studied through the lens of inequality especially in development context. The very idea of gender was further developed

R. Banerjee, G. Samanta, *Negotiating Terrain in Local Governance*, Local and Urban Governance, https://doi.org/10.1007/978-3-030-60663-3_1

in the 1980s under the theoretical influence of Marxism and Socialist Feminism. At that time, the third wave of the movement labelled post-feminism reflected the debates started by women of colour and lesbians focusing on politics of location, oppression and common identities. This also led to the development of post-feminism (1980s onwards) that dealt with microscale gender dimensions such as body, distance, place, space and imaginative environment (Johnston 1997; Peet 2004; Dikshit 2008).

Over the decade of the 1980s, women from both developed and developing nations developed solidarity despite ideological and cultural differences through their participation in different world conferences. The Fourth World Conference on Women was held in 1995, in Beijing, for action concerning human rights for women. The platform was developed earlier, with the development of a women's movement in the third world. These movements were not necessarily for feminist agendas but to promote feminist perspectives, for example, the Chipko Movement in India, Green Belt Movement in Kenya, Mother of the Disappeared Movement in Latin America and so on. Increasingly, women's groups around the world united around issues of economic justice, human rights and degradation of environment with the idea of being united in diversity (Johnston 1997; Peet 2004; Dikshit 2008).

However, women still remain invisible in different spaces, especially in public spaces. For example, if we observe the data of women's workforce participation in working space, it shows a decreasing trend. Across the globe, the gender gap is one of the most pressing challenges facing the workforce. The International Labour Organization (ILO 2017) 'World Employment and Social Outlook (WESO): Trends for Women 2017' report states, 'Women are insignificant, likely than men to participate in the labour market, and those who do look for work are insignificant, likely than men to find it. What is more, women in the [south Asia] region work in jobs that tend to be both of lower quality and lower pay than their men counterparts'. The reasons for such low quality and lower pay to women include lack of affordable childcare, the need to contribute time to household chores and social expectations from women. In India, for instance, women tend to be segregated into sectors of the economy that are not growing, resulting in 20.7 million fewer women employed over the period from 1994 to 2010, ILO research has found.

The report finds that over 50% of the women polled in South Asia would rather be working in a paid job than staying solely at home. According to the International Labour Organization report (2017), the reasons include (a) imbalance in work–life and family–life; (b) disapproval of family members; (c) abuse and discrimination ranked as the top reasons holding them back; and (d) organized sectors mainly sanctioned special leave during the pre- and post-natal period of motherhood to the women employees. However, women who work in the unorganized sectors cannot avail such facilities, and that is why their participation stops after a certain phase of their lives.

Challenging these social norms—including issues related to discrimination, education, unpaid care work, work–family balance and marital status—will be key to engage more women in the workforce, the WESO report says (Rukmini 2017).

In the government system, women are still struggling for their political opinions to be voiced, even after a century of receiving the right to vote. At present, even in cities, they are being continuously excluded from formal government systems, and their concerns rank poorly in the listing of urban priorities. This is despite the fact that a number of countries have passed ground-breaking legislations facilitating women's participation in the decision-making processes of local governance in order to increase their numbers.

Phillips (2003) in the book *The Politics of Presence* explained the concepts of 'politics of ideas' (documented reserved seats) and 'politics of presence' (actual representation of women); further, Pitkin's theory of representation (1967), that is, formal, descriptive, symbolic and substantive, is important to mention (Bharathy 2012; Nakkeeran 2012). Both these concepts represent the distinction between numerical and effective representation of women and the impact on governance. At present, in cities, the 'politics of ideas' is given more emphasis than 'politics of presence', and women continue to be excluded from the formal structures of governance.

In Pitkin's theory, symbolic representation focuses on the representation of women by reservation, and substantive representation analyses the response of elected women representatives in political decision-making (Nakkeeran 2012). Thus, it is clear that the issue of women's representation goes beyond the numbers and urban women continue to be excluded from decision-making processes (CMAO 2009: 1). This, however, does not mean that women are not worthy of doing so.

If we go through the political history of different countries of the world, we notice examples of many powerful ladies in recent politics. For example, South Asia has seen the rise of more eminent women in politics than the rest of the world, such as Chandrika Kumaratunga in Sri Lanka, Sonia Gandhi in India, Benazir Bhutto in Pakistan and Khaleda Zia in Bangladesh (Omvedt 2005: 4746). There is yet another iron lady by the name of Margaret Thatcher who developed new economic policies in the political space of Britain. If we observe the recent women's leadership throughout the world, Theresa May in Britain and Angela Merkel in Germany are women who have proved their leadership qualities in the developed nations of the world. Hillary Clinton in America and Sonia Gandhi in India have also proved their leadership qualities.

As for developing nations, Nepal has a woman President, Bidhya Devi Bhandari (2015 to present). Aung San Suu Kyi led the National League for Democracy (NLD) to a majority win in Myanmar's first openly contested election in 25 years in November 2015 and became the President of the country. South Korea elected their first woman President named Park Geun-Hye in 2013. In Taiwan, Tsai Ing-wen, and in New Zealand, Helen Clark are women Presidents who have developed their nations gradually. Besides these, if we go through statistics, we observe that though women represent half of the world's population, they are underrepresented in all political positions. For example, in June 2017, the global average of women's representation in Parliament was only 23.3%, rising up from 20.3% in 2012 (IPU 2017).

Considering South Asia as a good example where one can observe powerful women politicians, analysis shows that overall women's political representation is dismal in nature (Asia's regional average was just 19.5% in 2016; see IPU 2017). The membership of women in national parliaments across South Asian countries ranges from 6 to 8% during the last few decades, and most of them come from elite class backgrounds (Omvedt 2005: 4746). The percentage of women in national parliaments of South Asian countries was 5.9% for Nepal (1999), 21.1% for Pakistan (2002), 4.4% for Sri Lanka (2001), 8.3% for India (2004), 2.0% for Bangladesh (2001) and 9.33% in Bhutan (as per Online Women In Politics, 2005, and Global Database, 2005; quoted in Omvedt 2005: 4750). In recent times, the proportion of women members in political parties has been very low (7% on an average), and their representation in parliaments varies from 3 to 34% across countries. At the individual level, Bangladesh has 18.5% representation in the Parliament, Bhutan 8.5% in the National Assembly, Sri Lanka 5.3% in the Parliament and Nepal 30% in Parliament (ICRW n.d.; World Bank 2011).

From this, it can be proved that the gap between men and women is greater in the political space than other social spaces (Norris and Inglehart 2000, quoted in Beaman et al. 2006: 1). This gender gap in political representation coexists with a well-established gender gap in political views, indicating that women and men have different policy preferences, with women more likely to support policies particularly spending on childcare and other child-related expenses (Lott and Kenny 1999; Edlund and Pande 2001; Edlund et al. 2003; all quoted in Beaman et al. 2006: 1). Political reservations for women are often proposed as a way to enhance women's ability to participate in policy-making rapidly. Quotas for women in assemblies or in party candidate lists are enforced in the legislation of over 30 countries (World Bank 2001; quoted in Chattopadhyay and Duflo 2001/2003: 1) and in the internal rules of at least 1 party in 12 countries of the European Union (Norris 2001; quoted in Chattopadhyay and Duflo 2001/2003: 1). However, the fact represents 'politics of diversity' in different countries as well as different levels of governance.

Among the Asian countries, if we go through the case of India, we observe that the largest democracy in the world, India, elected its first woman President in its 60th year of independence. This clearly reflects the position of women in Indian governance (Sarkar 2015: 194–195). The 73rd and 74th amendments for Panchayats and Municipal bodies give women a way to move forward in their political career. However, despite the growth from 4.4% in 1952 to 11% in 2009, their representation is still below the world average of 20% (Rai 2011). Recent statistics reveal that India lags behind many countries, including the neighbouring countries of Pakistan and Nepal, when it comes to women's participation in politics. With only 11.8% women's representation in the Lok Sabha and 11.1% in the Rajya Sabha, India ranks 98th in the world, according to data released by the Inter-Parliamentary Union (IPU), an international group that works for promoting democracy, peace and cooperation in the world (IPU 2017).

Moreover, if we look closely to the scenario at the state level (Rao 2019), we find that women Members of Parliament (MPs) from West Bengal only cover 12.5% in

Lok Sabha (2014). While the numbers have increased over the years, the percentage of women in the Lok Sabha even today is not a figure to be proud of, especially when the ideal number should be 33% at least. In case of Rajya Sabha, the number is far more insignificant, that is, only 9% (ADR n.d.).

The Women's Reservation Bill or the Constitution (108 Amendment) Bill, 2008, which proposed to reserve one-third of all seats in the lower house of Parliament and of State Assemblies, has never been passed. It was last passed in the Rajya Sabha in 2010, but eventually lapsed due to lack of consensus in the Lok Sabha. The National Democratic Alliance (NDA) government during their tenure since 2014 has not tabled any version of the Women's Reservation Bill, even though it had been part of the assurances made by the BJP (Bharatiya Janata Party) election manifesto in 2014. They had a commitment of 33% representation of women in Parliament and State Assemblies before election.

At the level of local government, the reservation of seats for women is 33%, according to the 73rd and 74th Constitutional Amendment, 1993. Recently, the issue of 50% reservation of seats for women in local bodies has been discussed. The Constitution Bill of 2009 (110th Amendment and 112th Amendment), to reserve half of the seats in all rural and urban local bodies for women, was first introduced in Parliament by the United Progressive Alliance (UPA-II).[1] However, it lapsed with the dissolution of the 15th Lok Sabha. Over the years, many states of India such as Andhra Pradesh, Assam, Bihar, Chhattisgarh, Himachal Pradesh, Jharkhand, Karnataka, Kerala, Madhya Pradesh, Maharashtra, Odisha, Rajasthan, Tamil Nadu and Tripura have increased this quota to 50% in Panchayati Raj institutions, i.e. rural local government. Many states, including West Bengal, have also supported the proposal for providing 50% reservation for women in the ULBs (urban local bodies). Thus in India, women are better represented in local-level governance due to the reservation, and the higher levels of government such as State Assemblies and Parliament do not have many women, as there is still no reservation for women.

Therefore, the main causes of lesser representation of women at the higher level of political system are as follows. At the level of local government, women representatives, while facing a multitude of social and administrative problems, have to prove their capabilities. According to Ghosh and Lama-Rewal (2005), there is a wrong notion harboured by citizens regarding the efficiency of women representatives, as it is believed that women are under greater pressure to perform in comparison with men. However, it is worth noting that during discussions about the issue of women's reservation, there is no question about women's capability, but the focus is on equal rights for women. Therefore, the main purpose of reservation is to improve the status of women in governance as well as in society. Again the vital questions are: Does reservation for women really provide equal opportunities to them to represent their priorities in decision-making? Is there any discrimination related to

[1]The **United Progressive Alliance (UPA)** is a coalition of centre-left political parties in India formed after the 2004 general elections. The largest member party of the UPA is the Indian National Congress. The second tenure of this government which started from 2009 to 2014 is called as UPA II Government.

gender justice? Does their family provide proper support to them? What role do political parties play in the emergence of women leaders at the local level? Are women able to adapt themselves to their changing socio-cultural life due to their political participation in governance?

To answer these questions, many studies have been conducted all over India. If we look at the studies conducted in different parts of India, we observe that most of those studies are focused on the Panchayat level, or rural local government, than the urban local bodies. Thus, the existing literature on the impact of the 74th Constitutional Amendment on women in running city governments is meagre in proportion. The reason may be that a majority of the Indian population lives in rural areas and village-level social relations are very complex, thus giving rise to an unconscious interest in the Panchayati Raj (Ghosh 2011: 13). In case of urban governance, whatever research exists is mainly focused on the metropolitan cities, and small cities remain under-researched. The present research thus highlights the context of women's representation in the cities. The following section highlights the review of existing literature on issues linked to women's participation in local government, with a special emphasis on women in urban local governance.

1.2 Women's Participation in Politics and Governance

The term 'participation' is widely used in different ways in different fields. Participation in a psychological sense may be thought of as interaction between people to achieve specific goals or outcomes. It is not a unitary concept and consists of interrelated elements, which may be manifested in the decision-making processes of an organization in diverse ways. According to Singla (2011), there are three elements to the concept of participation: influence, interaction and information sharing. Thus, effective participation will ultimately bring empowerment. However, there are many obstacles in the way of women's participation in governance.

The major obstacles hindering the participation of women in the governance of India have been identified by other scholars (Omvedt 2005; Kodoth and Mishra 2011; Rai 2011; Ghosh 2010; Singaray 2014) as follows: (a) families and society; (b) patriarchy with class; (c) caste–gender subordination; (d) placing women in less important ministries such as women and child welfare, cultural department, etc.; (d) dowry, treating women as burden and conservative religious views regarding women; (e) nature of political parties, for instance, the communist party has no woman in high-level politics; and (f) politics as the source of power and income, which only men attempt to control. We can categorize these factors under some broader headings:

(a) Ideological factors such as patriarchy, which has placed women within the private domain, that is, home and family, are considered as a primary training ground for women who aspire for a career in politics (Bari 2005: 03; Nair 2012: 481; Sundstrom 2008: 18).

(b) Political factors such as men having dominated politics and political parties with formal political structures, and the fact that the political party is considered as an organ of political structure. The party makes the crucial decision about whether women should be allowed to contest from general wards or not and also determines the winning ability of a candidate (Bari 2005: 04; Nair 2012: 479). For these reasons, women have to take up less influential positions in order to make their position in the political parties through which the Indian democratic system operates (Mandal 1999; Gupta 2006).

(c) Socio-cultural factors such as managing time to maintain their dual role in public and private spheres, in light of social conservatism and rigidity (Sundstrom 2008: 18).

(d) Economic factors, such as lack of ownership of productive resources (Bari 2005: 05; Sathiabama 2012). Beside these factors, there are still others such as lack of protective security, coordination and networks, leadership-oriented training, insufficient media support, feelings of subordination towards women who are elected in reserved seats, proxy representation, backing, ignorance, ridiculing, interrupting, threatening and the inherent quality of women to remain non-vocal and invisible, institutional constraints (vote of no confidence for the women), discrimination in meetings and so on (Palanthurai et al. 2009: 60).

Women face problems related to decision-making because of the barriers inherent at the personal and at the municipal levels (Rizvi 2006). At the personal level, their lack of experience of municipal functioning, and the fact that their participation is forced by family members, with no concession in performance of domestic chores, social customs and taboos, adversely affect their participation in the affairs of urban governance. At the municipal level, the attitude and behaviour of their male counterparts hinder the active participation of women as they are being suppressed morally as well as psychologically (Rizvi 2006: 203). Also, women have to place their priorities according to the wishes of the political parties they represent, hindering their unity on issues related to women (Rizvi 2006: 204). It is important here to mention that education and reservation are not the only enabling factors of women's empowerment in politics; in addition, women also need a supportive family that preferably has a political background, someone in the family who is willing to take over their duties, which require them to care for the family and children, and a good amount of surplus money in order to participate in politics and governance (Srivastava 2015).

While considering the major hindrances women face in developed countries, Sheel (2003) found that Japanese women are also not able to participate in politics due to the following reasons: historical and cultural factors; industrialization and modernization relegating women largely to the private sphere; the '*ryosai kenbo*' (good wives and wise mothers) ideology that perpetuated into a web of entrenched values, attitudes and customs binding women to home and the family; new job opportunities; pro-women state measures as well as new technologies making women the 'most dynamic sector'; and so on (Sheel 2003: 4097). On the other hand, Horowitz (2009) observed in England that the major barrier for women is to manage their time for the work of councillor. For that reason, most of the women councillors in England are retired persons. The other obstacles to their participation include selector hostility, voter's preference towards men members, lack of resources and networks, lack of training and support and traditional beliefs regarding women (Morgan 2003; Horowitz 2009).

In South Africa, the basic requirement for a woman councillor is that she should be the watch dog to implement policies related to the needs of common people, with special reference to women. However, in reality, the system is dysfunctional in nature because of barriers such as excessive work load, collapse of municipal administration, poor channels of communication with public and officials, centrality in decision-making, dysfunctioning of ward committees, AIDS-related illness of councillors, lack of accountability of local government, predominance of ruling party councillors, lack of monitoring systems and non-responsiveness of municipal managers, among others (Paradza et al. 2010).

In the context of Asian countries, Khattak (1996) has looked into the condition of Pakistan and Nepal. In Pakistan, women remain invisible in the political domain because of lack of reservation, indirect election and illiteracy. In Nepal, although Dalit women are able to participate in the village-level politics, they can never participate in the decision-making process. The major decisions are usually taken by influential members (Khattak 1996: 10). In Bangladesh, the major hindrances to women's participation in decision-making in the political domain, as observed by Panday (2008), are proxy nomination of women candidates by political parties, strong Islamic ideology reducing the chances of women to participate openly in the political space, less mobility and inconvenient time of meetings to keep women away from taking part in decision-making as well as cultural barriers such as responsibilities of family, children and secondary obligations of social and political institutions, and so on.

Despite those enormous obstacles in different countries of the world including India, women's role in governance is not negligible. According to Panday (2008), the significance of women as representatives in the field of governance are as follows: (a) the different values, ideas and behaviour of the women shed new light on power distribution; (b) women can develop a new culture through their contacts; (c) it is a question of democracy and equality as well as a question of civil rights, thereby making the demand for proportional representation of women in politics unavoidable; (d) women's insignificant presence in politics raises questions about the legitimacy of the democratic process and of decision-making authorities; and (e) women are well informed and experienced about the basic problems and needs of the society. However, they are deprived of equal shares if they are not properly represented in politics.

There is also an argument that women bring new ideas, approaches and energy and put women's issues forward in public discussion and it helps to use the unutilized potential of women. They are also perceived as the pathfinders for attacking social issues with a different set of skills, leadership qualities and opinions (Joseph 2011). Empowerment of women in administration reduces gender conflict and enables gender-neutral policies. It is also a tool of economic, political and social mobility for women and encourages implementation of income-generating programmes, which would lead to the development of basic services in urban areas by women administrators (Kumar 2006). Women understand the difficulties of not getting proper services as they face it directly. For instance, women have to fetch water for their households if the water is not available. As a result, if they are in administration, they will make efforts for the betterment of such basic services. Women also understand women's problems better. Men are guided more by political motives as they consider it as a chance to enter state-level politics; however, women have the experience of the issues first-hand and thus, are more genuinely concerned (Upadhyay and Bhardwaj 2006). Women have also proved their excellence as team leaders as they have a deeper sense of loyalty and commitment as compared to men. Society has to decide how this intangible, but most valuable resource can be brought in practice (Narain and Mridubhashini 2006). Beall (1996) opined that in order to ensure gender-sensitive governance, there must be two objectives to look into—the first, to maximize women's participation in the political arena and, second, to foster gender awareness and competence both in men and women, so far as governance is concerned.

While analyzing women's role and participation, it is very important to mention that in an urban setting, women serve a larger share of population with far more complex problems than in a rural setting (Nair 2012). They also have the inherent constraints of having dual responsibilities at home and at their workplace. Due to all these adverse circumstances, they tend to take assigned jobs as challenges (Narain and Mridubhashini 2006). Kodoth and Mishra (2011) found that women councillors show greater courage in taking up development-related work, especially regarding issues of women's development and empowerment. However, he also observed that in some cases, they are guided by male representatives, especially when women are from lower castes and illiterate background. Ghosh (2010) had a different observation in her study on women councillors in Chennai. She observed that women councillors do not give priority to the issues of women and children. However, they are concerned about the issues of women and handicapped children. She also opined that this observation does not fulfil the agenda of women's reservation as a whole, due to the lack of awareness of women councillors on women's issues.

On the relations between women's political participation and crime against women, Iyer et al. (2011) found that it gives the women a voice when reporting in police stations. Moreover, women representatives from minority communities also help in articulating their grievances and lending a voice to their group members against crime. It certainly creates a positive impact and makes the police more responsive to crimes against women. Further, reporting biases are expected to be minimized, and women's exposure to potential crime situations as a function of the time they spend outside their home should be reduced. While considering the level of efficacy, women's participation in village-level politics is more effective in reducing criminal cases than at higher levels.

Kulkarni (1996) found that women's participation at the village-level is observed as a token in *Pani Samiti* meetings. Sometimes they simply do not know why they are selected. However, Beaman et al. (2006) showed that as water managers, women have the ability to take decisions regarding the proper management of water. Moreover, as good policy-makers, women give more priority to improving drinking water facilities, reducing the gender gap in school attendance and childcare facilities, taking action against bribes, raising voices for women, and so on. In spite of all these good activities, they often receive bad responses in meetings from the male leaders (Beaman et al. 2006: 18), which can be considered as resistance from patriarchal societies limiting women's empowerment.

The conditions of women as government representative vary widely depending on the specificities of space. Focusing on the space-specific studies available, we find that the role of women representatives and their concerns about issues varies with the nature of space and their specific needs. For example, women *pradhans* of West Bengal give more emphasis to the issue of drinking water facilities (Beaman et al. 2010: 19), while in Rajasthan the investment in roads is a more pertinent issue under the area of the Gram Panchayat having a woman *pradhan* (Chattopadhyay and Duflo 2001/2003: 3).

The space specificity has also been observed in a comparative study on the role of women councillors in Kolkata, Delhi, Mumbai and Hyderabad (Ruet and Lama-Rewal 2005). The study reveals that the work efficiency of women councillors varies, depending on the local context. In Kolkata, disinterest of officials and party influence determines the work of women councillors. In Hyderabad, lack of coordination and insignificant participation of the women councillors is prominent. However, in Delhi, women councillors are doing good work for the improvement of services. In Mumbai, the context is different as the people of Mumbai are more likely to go to corporate officials to solve their problems rather than to the local councillors.

The service delivery and democratic participation also vary with a gender-specific outlook as observed by Sathe et al. (2013). They observed that the service delivery and democratic participation among the villages having female *sarpanch*s are higher in the long run than those of villages having male *sarpanch*s.

The role of women councillors can also be judged by using the capability approach given by Amartya Sen. Vanishree Joseph (2011) has used this method in her study to evaluate the capability of women representatives at the grass-root level of governance in Tamil Nadu. She has classified the measures for capability based on two kinds of variables. One is freedom enjoyed by the women representatives, and the other is the functions they perform while having this freedom. This freedom has further been classified into five types: political, economic, social, transparency and protective security. The functions are also classified according to the local demands of the area, for example, administrative work, awareness regarding gender issues, providing services, problem-solving mechanisms, and works related to women. With these variables, the author has provided an insight to analyze the role of women councillors in the development activities related to governance in Tamil Nadu (Joseph 2011). Similar observations have been cited by Ghosh (2003).

In Chennai, women councillors are also aware of the issues of giving loans for generation of self-employment, pensions for the poor, road repairs and sanitation for the poor areas (Ghosh 2003: 133).

Community participation is considered a significantly important factor in empowering women councillors. In a study on South Africa, Paradza et al. (2010) analyzed the root causes of women's low performances in governance. They observed that the local government is dysfunctional and the administrative structure is the main focal point of anger and community protests. Residents are helpless and say that councillors never listen to them. On the contrary, the councillors also cite that they are unable to perceive the residents' problems due to weak public participation. However, the inherent cause behind it is the weakness in the capacity of the state to provide support to the local government. To mitigate this problem, women councillors have also taken initiatives such as encouraging and educating the residents to participate in the governance system (Paradza et al. 2010: 37).

1.3 Women's Seat Reservation[2] in Governments

Radical arguments assume women as an idol of ethnic care and compassion, whereas interest arguments treat them as a different interest group. In such a context, the justice argument has given more emphasis on universal justification of democratic practices. However, according to this study, the justice argument of reservation does not fit in case of India, because of caste–class differentiation and patriarchal ideology (Ghosh 2009).

In India, women's reservation process was initiated in the 1930s and 1960s. However, it was rejected in the constituent assembly as it was not supported by the principle of equity. In the first three elections, women were treated as a separate category (Deepmala 2012). After that, the parties realized that the issue of caste was more poignant than gender. In 1974, with the report of the 'status of women committee', reservation became an issue important to fostering equality. However, women's development became the focal issue here rather than their right to political power and representation. Then, with 'National Perspective Plan for Women', the 73rd and 74th Constitutional Amendments bill was passed in 1993, in which women's reservation for local government was ensured (Ghosh 2009: 16; Nair 2012: 470). The 73rd Constitutional Amendment was meant for Panchayats and the 74th for Municipal Corporations, Municipal Councils and Nagar Panchayats.

However, women's reservation became a matter of intense debate in Parliament among the political parties, which lead to the non-passing of the bill even after table up many times. The United Progressive Alliance and National Democratic Alliance

[2] Reservation of seats for women in the election of democratic governments is a strategic policy to increase the level of women's representation in governance. This policy is in practice in many countries especially in the countries of the global south, where women's representation is abysmally low in governance.

wanted to increase the total number of seats from 543 to 722 and then reserve 179 seats for women, whereas the Communist Party of India (Marxist) was in favour of introducing the bill in the original (1996) form (Ranjan and Retnakumar 2005: 4190). Ranjan and Retnakumar also opined that the basic cause behind such a controversy on the nature of the bill is the lack of clarity on the women's reservation bill proposed by different states in India.

Also, the long-term formula regarding the distribution of seats leads to no profit in short-term politics. Another problem is regarding the numerical observation. Lastly, the north–south debate regarding the 91st Amendment of the Constitution and being unable to refute the demands of Scheduled Caste (SC)/Scheduled Tribe (ST) representation as per Article 330 under the Constitution adds to the problem (Ranjan and Retnakumar 2005: 4191).

To overcome these problems, several measures have been suggested in the studies.

- Instead of reserving constituencies exclusively for women, it is suggested that it should be mandatory for every recognized political party to nominate women candidates for election to one-third of the constituencies by converting all the constituencies into dual member constituencies (Ranjan and Retnakumar 2005: 4191).
- Facilitating greater representation of groups without disrupting the present balance between different states (Ranjan and Retnakumar 2005: 4191).
- Waiting till 2026, when the existing freeze on Lok Sabha constituencies will end, let the population reach a stabilizing stage, only then increase the total size of the house and fix an appropriate quota for all deserving sections of the population (Ranjan and Retnakumar 2005: 4192).
- Conversion of 33% of Lok Sabha seats (179) into double member constituencies, so that election of one man and one woman each may be enacted (Ranjan and Retnakumar 2005: 4192).
- Promotion of genuine representatives of women at the lowest level and a representative structure to open up avenues for wider participation, articulation of women's concerns and the emergence of genuine leadership (Mazumder 1989: 2795).

This is the condition of women representatives and the scenario of reservation given to women candidates in India. But what happens to woman representatives outside India? Do they take their decisions independently? Do they enjoy any protection of representation through reservation of seats in the election process? We now examine the conditions of women representation outside India.

The studies of economic and political rights of women in Africa and Britain reveal that in African countries, women have received formal political rights before obtaining economic rights. However, women's inclusion in politics was achieved mainly after suffrage (the right to vote) and gender quotas (Doepke et al. 2012: 6.3). Doepke et al. (2012) have also found that when women are involved in politics both as voters and as policy-makers, the composition of government spending shifts towards higher expenditure related to health and children. Beall (1996) opined a similar argument that women's social responsibility towards their home also includes family healthcare, which is why when they are in governance, they give more priority to health.

Doepke et al. (2012) also stated their views on the political rights of women and argued that countries with higher income exhibit a larger share (21.3%) in political rights of women compared to the rest of the world. To explain the connection, they used the phases of development of women's rights in Britain and the USA to establish the previous argument. In England, during the second half of the nineteenth century, women received economic rights. During the twentieth century, political rights were extended to women. Finally, women received full equality in the labour market and legal rights. In the USA, women's suffrage movements began to emerge during the first half of the twentieth century. In Britain, women received partial suffrage through the People's Act of 1918. After the People's Act of 1928, women over the age of 21 got the right to vote.

In the USA, it has been observed (Doepke et al. 2012) that women's participation in politics was associated with a 24% increase in the state's social spending on local public health, which led to an 8–15% decrease in child mortality rate. Gender quotas have also been introduced to increase the presence of women in politics in some countries. Besides reservation, these countries have also taken some measures to provide a facilitating environment for women's participation in governance.

The Government of South Africa has taken several measures for the betterment of the work environment of women councillors. These include (a) capacity building courses for councillors; (b) taking photos and maintaining a record of unfinished work; (c) one-on-one meetings; (d) assessment of the work of councillors by respective political parties; (e) development of networking opportunities; (f) clear guidelines for councillors; (g) providing municipal transportation; (h) accommodations and equipment for councillors and ward committee meetings; (i) recommendations for fixed date for meetings; and (j) development of e-governance system, etc. (Paradza et al. 2010). After studying the performance of women councillors, this study recommends some other measures to be taken up for the betterment of the work experience of women councillors. These are regular assessment of councillors' performance, level of attendance, level and frequency of their contact with residents and citizens and so on. Paradza et al. (2010) opined that if these measures can be taken up, it would help the authorities to track the councillor's work and decisions, which ultimately lead to good governance.

Haug (2012) in his comparative study on Nepal and India has highlighted that in Nepal, 33% reservation of women's seats in the Constitution Assembly represented a significant improvement in 1999 along with a provision to elect one woman in village development committees. The country has also initiated gender-responsive budgeting (2007–2008) to improve gender sensibility. In comparison, in India, women's quotas were written afresh on a blank slate, as this was not as popular as caste quota. Here, the 73rd and 74th Constitution Bills were developed for 33% women's reservation in village as well as in municipal governance (Haug 2012), but never for the national government.

Among the studies of developed countries, Sheel (2003) has cited the gender-based measures for women's participation in governance of countries such as Japan and the UK. A law exists on gender-equal society of Japan, to encourage women's participation in decision-making bodies. In the UK, in London, local authorities instituted equal opportunity for women and minorities in the 1980s. Even earlier, the Maud Committee in 1967 had formulated a report based on the work and responsibilities of councillors in municipal administration.

Pakistan is a country that has an orthodox ideology regarding women. The Pakistan Government in 1979 passed an ordinance based on Article 32 for women's reservation. However, except some states no one bothered about it (Khattak 1996: 10). Therefore, Khattak (1996) suggested some additional measures to make the reservation a reality in Pakistan. The measures are (a) increase of women's quotas; (b) development of open and direct method of election; (c) political parties should be legally required to give a percentage of party tickets to women; (d) legal requirement of male and female chairpersons; (e) initiation of political education and training to women councillors; and (f) active role taken by non-governmental organizations (NGOs) and media to encourage women's political participation (IWRAW n.d.).

Panday (2008), in his study of women's participation in political decision-making in Bangladesh, suggested several improvement measures. These include (a) to realize the gender inequality, and to reorganize the roles of women in all sectors where TV, radio and newspapers can act as policy changers; (b) the government should take initiative for legal and constitutional reforms; and (c) political parties should take initiatives for gender equality.

Based on the review of available literature, the necessary requirements for women to be engaged in good governance can be summed up as follows: (a) making the system accountable and defining and organizing women's work within their specific social context in which they can operate (Stewart 2006); (b) increasing the political voice of women by increasing quotas, training, civil services, development of women-friendly institutions, promoting political awareness and giving party leadership to women (Baldez 2006); (c) administrative accountability by developing gender-sensitive complaint mechanisms, gender-responsive budgets, gender-sensitive design and implementation of programs and projects, gender-disaggregated and gender-sensitive monitoring indicators; and (d) reinforcement of client power by developing women's Self-Help Groups (SHGs) in governance, taking affirmative action in user group membership and gender-sensitive citizen monitoring and auditing (Horowitz 2009: 39–40).

Within India, reservation is not the only way to empower women councillors and to increase their capability in governance. Different studies undertaken by different scholars suggest several recommendations to make a visible space for women representatives in governance. First of all, Sekhson (2006) has mentioned the initiatives taken by Alochana (a non-profit research and documentation centre on women), which has developed a training kit regarding political education and experience for women leaders. Gupta (2006) has also supported the necessity of training for the women's representatives in politics. He opined that training is necessary not only during the pre-entry phase of women in governance but also in the post-entry phase, for developing leadership and communication skills to enable them to function effectively. Gupta (2006) also argued that women who are elected for the first time are educated but have little knowledge about working in the government, which is why training is required. In this context, Palanthurai (2012) in his study on Tamil Nadu focused on the recommendations for building capacity among women leaders to manage crowds. In Tamil Nadu, the Women Leader Federation was developed. It has also helped to make this 'critical mass' (women leaders) visible. Palanthurai (2012) has also highlighted the role of 'Gandhigram Institute of Experts and Development' for developing transparency among the councillors. The Institute of Experts have suggested four kinds of literacy (money, body, civil, general) among the councillors, which can be developed through training and regular interactions. Gupta (2006) has also suggested that women should also be trained in the critical areas of urban governance such as environment, health, water, conversation, education and sanitation, and they should be given sufficient financial resources to solve the problems of the people in their areas.

A similar argument has also been cited by Raj (2012). He suggested that women leaders also have to be trained in order to enhance their sense of duty to take action on women centric issues. He cited the example of 'Kerala Institute of Local Administration' which provides training and gender awareness classes to women leaders. The institute has also developed a forum of political parties and initiated gender-responsive budgeting. As a result, women representatives have taken initiatives such as component planning for women, formulating projects for women, and developing vigilance cells such as Jagratha Samithi, Anganwadis, Kudumbashree Mission, etc. (Raj 2012). Holzner and Wit (2003) have also cited the example of Mumbai Government for the arrangement of leadership training of women leaders in collaboration with the Netherlands. Wit has also observed that after the training, women leaders became more vocal and worked on par with their male counterparts in order to become transformative leaders.

CMAO (2009) of Orissa has also developed a learning kit for women councillors. This kit is concerned with the development of knowledge of the women councillors about their responsibilities and how to dominate and put forth their viewpoints in meetings. It also focuses on media exposure, awareness about updates and so on. Drage (2001) has suggested strategies for the improvement of the work of women leaders in local government in the countries of Asia-Pacific, including increase of salary for councillors, funds for campaigns, development of networks through the legal system, training related to their responsibilities and so on.

Despite careful deliberation and measures taken up in different countries, the issue of proxy representation of women in governance still needs to be addressed.

1.3.1 Rethinking Proxies

In the discourse on women's reservation in governance, it is often argued that elected women act as proxies or place-holders for men. Ghosh and Lama-Rewal (2005) and John (2012) have explained the concept of proxies in detail. This may be either in case of seats or in case of their ability to do work. In the first group of proxies, women are elected to the proxy seat of the men because of reservations, to secure the seat of the representative for the male members of the next generation. In the second group of proxies, women share their work with the male members of their family. Nair (2012) has stated some observations about the second group of proxies. These women attend meetings and continue their work in the home and in office; however, ward work is done by their husband, brother or son.

The reason for proxies among women representatives has been highlighted by Singh (2006). He has identified three reasons: (a) negative influence of family members (involvement in political work); (b) the nature of political parties and their mercy; and (c) opposition of the political parties (if they belong to the opposition group they cannot perform their role) (Singh 2006: 61–62). Palanthurai (2012) has also argued the reasons of proxies of women leaders. Firstly, the men of the elected women representatives' families attend meetings, carry out their activities, claim the cheques and act like officials themselves, and the other reason is that women representatives are not well-versed with corrupt practices and practice it through the male members of the family. According to him, this also makes the elected women proxies.

The question now arises, how to prevent proxies from being elected. One response to this has been to restrict the participation of women who are directly related to a serving politician. However, electoral politics is centred on such networks. In contradiction, Mishra (2015) has stated that the very idea of proxies of women representatives is unrealistic, as tokenism not only works for women but also for men as well, as they also depend on political leaders for re-nomination. Moreover, men obtain their position through this network, though they can use their position as they wish within the party limits. He also suggested that to avoid such double standards in case of women, women will have to think for themselves about their duties and responsibilities.

In this context, John (2012) has argued that there is a common assumption that proxy women are passive place-holders for men, but instead a whole spectrum of possibilities emerges on analysis. There are, no doubt, a few who remain victim figures throughout their term. However, there were also many women who developed in different ways in the course of their 5 years in office, some becoming significant political players in their own right. Joseph (2011) shared a similar argument that women attain a position that gives them power to do the work in their own way. Otherwise, women's presence simply reinforces the male-dominant structures in governance.

An emerging debate thus arises on the subject of proxies, where people's perceptions on proxies become very significant. Thus, in the overall assessment of women's position in governance, perception of citizens is very important to identify. Whether there is a proxy or not can be judged through the people's perception on their activity and efficiency. The next section deals with the literature on how citizens perceive their women leaders.

1.4 Women's Agency: Freedom and Functioning Principle

Women's emancipation and empowerment are difficult to judge quantitatively alone as these are, in most cases, based on social conditions. Thus, the research uses the application of the concept of 'capability' as given by Dr Amartya Sen (2000), as the dominant framework to analyse the role of women councillors in urban governance and measure this both quantitatively and qualitatively. This concept is based on two kinds of variables: freedom and functioning. By using these two variables, one can judge the capability, that is, being and doing of a person. It is also a reflection of their freedom to achieve valuable functioning.

The concept of freedom was first developed by Jean Paul Sartre in his philosophical essay, '*Being and Nothingness*' (1943). It was published in English in 1956 by Routledge publication. He distinguished between two types of freedom. One is the freedom 'to obtain' what one has wished; and the other is the freedom 'by oneself to determine oneself to wish'. The former can be termed 'freedom to obtain' and the latter 'freedom to choose'. 'Freedom to obtain' refers to our ability to act in certain ways in the practical world. 'Freedom to choose' refers to the fundamental projects that we set for ourselves (Nayak 1998; Franchi (n.d.); Iddo 2012). In this context, Sartre presents an example of a mountain climber who has the choice to climb the mountain. However, the success and failure depend on the situations and conditions. That success and failure are obtained by the climber to preserve his/her identity. According to him, freedom is absolute; he asserts that we are 'totally free' and that there is no obstacle in the absolute sense, but the obstacle reveals itself as a coefficient of adversity across freely invented and freely acquired techniques.

After Sartre, the concept of freedom was first popularized by Isaiah Berlin's work *Two Concepts of Liberty* (Yupanqui 2011). He explained the concept in two ways—one is positive freedom and another is negative freedom. Here, 'negative freedom' is the 'freedom from' negative factors such as harassment and assault, which are violating our social security. On the other hand, 'positive freedom' refers to having control over one's own life, that is, 'freedom to' (Garrett 2005; Yupanqui 2011). It also depends on two major factors: real self and ideal self. The real self encourages us to fulfil our wishes at any cost, whereas the ideal self helps to make a rational decision about our wishes. On the basis of that, positive freedom encourages us to be free to do what is right or rational. To remove any confusion between positive and negative freedom, MacCallun (1967) has given 'one concept of freedom' (as cited in Yupanqui 2011: 36). According to him, freedom signifies both the concept of positive and negative freedom. Thus, freedom is both, being free from something and to be or to do something.

Amartya Sen (1999) made the concept of freedom broader. In his book, *Development as Freedom*, he placed more emphasis on individual freedom (Sen 2000). Capabilities are the central idea of Sen's notion of freedom. The capability is the ability of a person to perform a valuable function. Such ability depends on his/her individual freedom. According to Sen (2000), freedom can be reflected in the person's capability, which is provided by both processes and opportunities. In light of this, expanding capabilities is expanding freedom. However, Sen has judged the freedom in two ways—primary end and principal means of development. Individual freedom becomes both the means and ends of development. The means of development denotes instrumental freedom which has a consequential effect of bringing about other freedoms or development of society. Primary end signifies substantial freedom, the substantial freedom provided by the society to achieve valuable functioning of an individual, that is, the implementation of instrumental freedoms (Sen 1993, 1999; Garrett 2005). In Sen's (2000) perspective, substantial freedom is what the instrumental freedom is supposed to bring about, in particular, justice and equality. However, achieving instrumental freedoms does not guarantee the achievement of substantial freedom (end of development) as the implementation depends on society.

Instrumental freedoms can be of many kinds, such as political, socio-economic, transparency and protective security (Sen 2000). The functioning is included in the functions performed by an individual having these freedoms. The ability to do functions depends on the freedom of an individual. Therefore, to achieve valuable functions, freedom to do that function is needed. For this reason, freedom is considered here as an independent variable and functioning is the dependent variable. The present research analyses the nature of freedom of the WCs (women councillors) which influences their functioning for the development of wards.

1.5 Having, Loving and Being Principles and the Quality of Life

'Having, Loving, Being: An Alternative to the Swedish Model of Welfare' was a concept given by Erik Allardt in the book, *The Quality of Life*, edited by Martha C. Nussbaum and Amartya Sen (1993). This concept is based on three main features: having, loving and being. Among these, 'having' refers to the material conditions which are necessary for the survival of an individual (economic and social condition). 'Loving' is related to attachments developed by a person with the people of his/her personal as well as professional space to maintain individual social identity. 'Being' stands for the participation of a person in decision-making and political activities. The detailed concept is explained in Table 1.1.

Allardt (1993) specified the parameters under different attributes. 'Having' is related to the perception of the individuals about their economic resources, housing condition, employment, working conditions (physical work, routine, measures of

Table 1.1 Variables of having, loving and being principle

Variables	Objective	Subjective
Having (material and impersonal needs)	Objective measures of the level of living and environmental conditions	Subjective feelings of dissatisfaction/satisfaction with living conditions
Loving (social needs)	Objective measures of relationships to other people	Unhappiness/happiness subjective feelings about social relations
Being (needs for personal growth)	Objective measures of people's relation to (a) society (b) nature	Subjective feelings of alienation/ person growth

Source: Allardt (1993)

stress, etc.), health and education. 'Loving' is related to attachment and contacts with the family, friends and local community and with other organizations. 'Being' stands for the extent to which the individual participates in decision-making activities, political activities, leisure-time activities, enjoyment and meaningful activities.

1.6 Central Ideas of the Book

The central idea of this book revolves around a central question: 'How are women's social and political spaces changing due to their political participation in the urban local governance?' To explore the answer to this question, we further devised a number of objectives, such as understanding the identity, socio-economic position and the conditions under which women participate in local governance as councillors in cities; identifying the level of capability (freedom and functioning) of the women representatives in different spaces (family, society and workspace); exploring the major problems encountered by these women in performing their responsibility both in private and in public spaces; analysing the capability of the women councillors in comparison with the work of men councillors in urban governance; evaluating the councillors' capabilities by the citizens; and understanding how women councillors' freedom and functioning vary over cities, and exploring the reasons underlying those differences.

The book is based on empirical research carried out in four cities of West Bengal, a provincial state of Eastern India. The cities selected are Darjiling, Balurghat, Raniganj and Hugli-Chinsurah. We selected West Bengal as a state to do our field research due to the fact that we understand the history, geography, culture and language of the state much better than any other states of India. Our personal experience helped us to bring subjectivity in selection. The local language also plays an important role in grounded research with qualitative and ethnographic methods, as in India each state has a different language. To understand the gender question much better, we need to depend on our local language and vocabulary to make conversation with the research participants without using a translator.

West Bengal, located in the eastern part of India, has a total area of 88,752 sq. km with 19 districts and three subdivisions (Fig. 1.1). Each district is divided into a

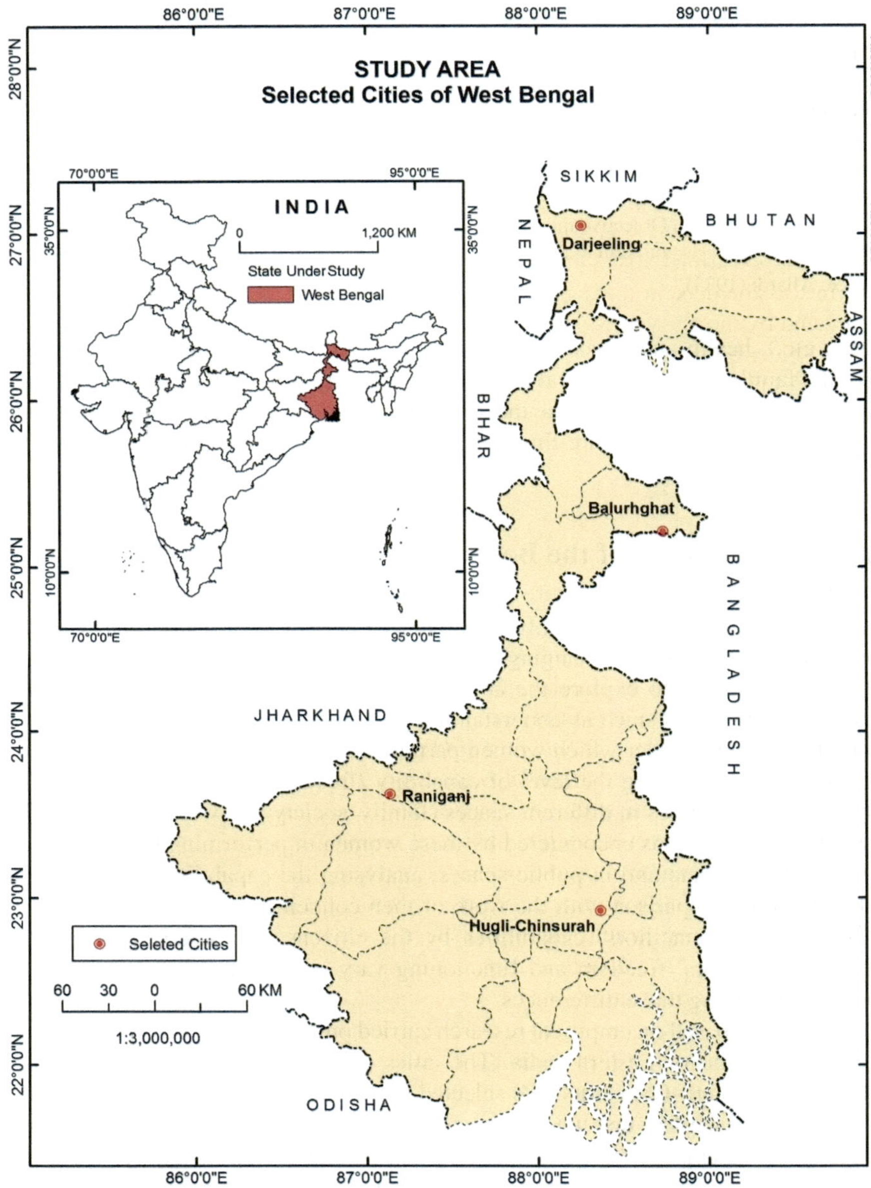

Fig. 1.1 Location of study area

number of community development blocks; and these blocks consist of Gram Panchayats (rural) and municipalities (urban). According to 2011 Census data, West Bengal has a total population of 91.35 million with population density 1028 persons per sq. km. In terms of political environment, it has a history of being led for 34 years by the communist party. However, since 2011, the ruling party of the state has been the All India Trinamool Congress party. Traditionally, local governance in West Bengal has always been under the control of the ruling party in the state with a few exceptions. Only a handful of municipalities and municipal corporations are under the opposition party. This seemingly minor information hints at a much larger doctrine, as the decentralization in the state is somewhat controlled by centralized policy and by the party in power in the state.

The selection of the four districts within the state of West Bengal was made on the basis of three parameters: (a) district-wise urban intensity index (Appendix 1.1); (b) city-level difference between the observed (number of women councillors elected) and expected (number of seats reserved for women) representation of women; and (c) the location of the cities in different socio-cultural regions of the state. The selection has been done in different steps.

First we calculated the urban intensity index. We grouped all the districts into four categories depending on their urban intensity index. Subsequently, the difference between their observed (number of women councillors elected) and expected (number of seats reserved for women) representation was also computed. This helped us to examine the acceptance of the women councillors to be elected in unreserved seats as well. With greater number of women elected in unreserved seats, the difference between observed and expected value increases, and if not, then there is no difference between observed and expected representation. Finally, the district-wise representation of women councillors to total councillors has been calculated (in percentage) on the basis of the available data provided by the West Bengal State Election Commission. This data helped in finding the quantitative strength of participation of women in urban governance. On the basis of the percentage representation of women councillors, we categorized the districts into four categories and selected one district from each category. The selected districts are Darjiling, Bardhaman, Hugli and Dakshin Dinajpur.

After the selection of districts, the following factors have been considered for the selection of cities within the selected districts. The cities considered for selection are small cities, with a population of less than 200,000. In addition to the population size of the cities, the nature of municipal board was also considered; in other words, the party in power and the percentage of women councillors (WCs) in the municipal boards were considered for the selection of cities.

From this analysis, Darjiling from Darjiling District, Balurghat from Dakshin Dinajpur, Raniganj from Bardhaman and Hugli-Chinsurah from Hugli District have been selected. Among these four cities, three are headquarters of three districts—headquarter of Darjiling District is Darjiling, headquarter of Dakshin Dinajpur is Balurghat and Hugli-Chinsurah is the headquarter of Hugli District. Raniganj in Bardhaman District has been selected because of its location near the border with Jharkhand state and in the midst of a mining and industrial area having a mixed socio-cultural context.

1.6.1 The Cities

Four cities were selected from among the small cities from four socio-cultural regions of West Bengal for an in-depth understanding of the socio-cultural phenomena affecting women councillors in that context in different small cities in West Bengal. The conditions of women in small cities are more closely tied with socio-cultural norms in comparison with the conditions of women living in metropolitan cities. That is why we targeted small cities for understanding women councillor's differential levels of emancipation, and their linkage with the local socio-cultural conditions (Table 1.2).

Darjiling city is located in the northern Himalayan region. The population size of the city was 1,18,805 in 2011. It has 32 wards with an area covering 10.57 sq. km. Among 32 wards, 16 were headed by women representatives. Balurghat is located in the Dakshin Dinajpur District and had a population size of 1,51,299 in 2011. The city government of Balurghat is headed by a woman, Smt. Sucheta Biswas as the Chairperson. She is a school teacher and has been an active member of the Revolutionary Socialist Party (RSP) since she was young college student. Balurghat has 23 wards and covers an area of 10.76 sq. km. Among 23 wards, 8 wards have women councillors.

Raniganj is located in the southern part of Bengal and in Paschim Bardhaman district and covers an area of 24.09 sq. km. The population was about 1,29,804 in 2011. Out of 22 wards, 11 wards had women councillors. The municipal government was under the Communist Party of India (Marxist) [CPI (M)] with Mr. Arup Kumar Mitra, a retired school teacher as the Chairperson. Further south, we selected Hugli-Chinsurah city, located in the Kolkata Metropolitan Area (KMA). It is located in Hugli District and covers an area of 17.27 sq. km. The population size was 1,80,502 in 2011. It has 30 wards of which 11 wards had women councillors. The then municipal board was governed by the All India Trinamool Congress (AITC) with Sri Gourikanta Mukhopadhyay as the Chairperson, an advocate by profession.

Table 1.2 Characteristics of selected cities in West Bengal

Cities	Area (sq. km)	Population	Population density per (sq.km)	FMR	% Female literacy	Gender disparity in literacy	% Female WPR	% Representation of women councillors
Darjiling	10.57	118,805	11,240	1007	48.92	2.14	28.51	46.87
Balurghat	10.76	151,299	14,061	998	48.63	2.73	24.31	34.78
Raniganj	24.09	129,804	5388	915	43.74	12.51	12.57	45.45
Hugli-Chinsurah	17.27	180,502	10,451	995	48.42	3.16	22.71	36.66

Data Source: Census of India and Office of the Municipalities

1.7 Research Process

The book is based on research including the triangulation between the two approaches—feminist and pragmatic approach—and thus includes both quantitative and qualitative aspects of research. Methodologically, the research depends on a post-positivist approach following Ryan (2006), as it is highly applicable to the present research. As a social research, it does not deal with the development of scientific models and does not keep the research away from human values. It deals with both theory and practice to analyze the complex relationship of human society, which cannot be explained by universal and generalized scientific models. Model building is usually targeted to universal applicability. However, this research is based on the ideology that the worlds (works and lives) of women councillors are complex and are highly dependent on the specific local contexts, which cannot be explained through scientific models.

As the gender question does not follow a linear trajectory, the present study emphasizes on both subjectivity and objectivity to understand whether the awareness, experience and knowledge of the person are multiple, relational and complex. The research focuses on problem setting rather than problem-solving, as there is no universal solution to the problems addressed. The social construction of narratives is an essential aspect of the research as it takes the reader into the story, and we strongly believe in the idea that narratives are not just the words, but a tempo to understand the social world.

For the field research, we targeted three groups of research participants—women councillors, men councillors and the citizens. The sampling techniques also varied with the target groups. As the number of women councillors is not very high, we covered all them except for a few who did not want to participate in the research process. The total number of women councillors interviewed is 38. For the selection of men councillors, the spatial stratified random sampling method has been followed. The sample is 50% of the total men councillors of the individual cities (Table 1.3). Total number of men councillors interviewed is 27. The location of their respective wards (centre and the periphery of the city) along with their position in the municipal government was prioritized in selecting the men councillors. For the selection of citizens, the sampling technique used is purposive. The total number of citizens covered for this survey is 400 including 100 in each city which roughly represents 0.1% of the total population in the selected cities.

Table 1.3 Number of research participants

	Research participants		
Cities	Women councillors	Men councillors	Citizens
Darjiling	14	7	100
Balurghat	7	6	100
Raniganj	8	6	100
Hugli-Chinsurah	9	8	100

The field research includes both quantitative and qualitative methods. The quantitative methods include tools such as semi-structured questionnaires, time-use diary and analysis of both primary and secondary data collected from various sources. The qualitative methods include participant observation, in-depth interviews and focus group discussion and perception survey.

For the selection of study areas, the secondary data of number of towns and the number of urban population of the districts of West Bengal have been analyzed. District-wise elected women representatives in terms of the seats reserved for them have also been analyzed from the data of West Bengal State Election Commission (2010–2013). The demographic information and information about the political identity of the councillors of the study areas have been collected from the websites of Census of India (2011) and of Municipal Affairs Department, West Bengal, respectively.

To know the profile of the women and men councillors, the secondary information has been collected from the Municipal office. To get the primary information about the women councillors, a semi-structured questionnaire survey besides in-depth interview of the women councillors has been implemented. The questionnaires have questions based on having, loving and being principles[3] and on the capability[4] (Freedom and Functioning) of the women councillors. The survey of the men councillors is also followed by a semi-structured questionnaire. Group discussions were organized at both the office premises and at the houses of some of the councillors. We also attended a number of meetings of the councillors of all the four cities as non-participant observers.

To analyse the time-use pattern of the women councillors, a time-use diary was developed and filled up in person to record the time they use for SNA, extended-SNA and non-SNA activities. The perception survey and semi-structured questionnaires were also used to identify the political awareness of the citizens and the perception of the citizens about their men and women councillors.

References

ADR (n.d.) Women representation among elected representatives. Association for Democratic Reforms, New Delhi. https://www.adrindia.org/download/file/fid/4977. Accessed on 16 May 2017

Allardt E (1993) Having, loving, being: An alternative to the Swedish Model of Welfare Research. In: Nussbaum M, Sen A (eds) The quality of life. Oxford University Press, New Delhi, pp 88–94. https://www.oxfordscholarship.com/view/10.1093/0198287976.001.0001/acprof-9780198287971-chapter-8. Accessed on 10 May 2020

---

[3] 'Having, Loving and Being' principles are a major concept of this research. This is discussed in greater detail in Chaps. 2 and 4.

[4] The concept of capability is based on freedom and functioning behaviour of the individual. It is a major concept of the present research and is discussed in detail in Chaps. 2 and 5.

Baldez L (2006) The pros and cons of gender quota laws: What happens when you kick men out and let women in? Politics and Gender 2(1):101–128. https://doi.org/10.1017/S1743923X06061010

Bari F (2005) Women's political participation: Issues and challenges (Draft). United Nations, Division for the Advancement of Women (DAW), Bangkok. www.un.org/womenwatch/daw/.../EGM-WPD-EE-2005-EP.12%20%20draft%20F.pdf. Accessed on 13 Dec 2012

Beall, J. (1996). *Urban governance: Why gender matters?* https://www.gdrc.org/u-gov/doc-whygendermatters.html. Accessed on 30 Jan 2014.

Beaman L, Duflo E, Pande R, Topalova P (2006) Women politicians, gender bias and policy making in rural India, State of the world's children 2007, background paper, UNICEF. UNICEF, New York, pp 1–23. https://www.unicef.org/sowc07/docs/beaman_duflo_pande_topalova.pdf. Accessed on 20 June 2013

Beaman L, Duflo E, Pande R, Topalova P (2010) Political reservation and substantive representation: Evidence from Indian village councils. In: India Policy Forum 2010|11. National Council of Applied Economic Research/Brookings Institution, New Delhi/Washington DC, pp 159–191. scholar.harvard.edu/files/rpande/.../politicalreservationandsubstantiverepresentation.pdf. Accessed on 26 Sept 2012

Bharathy M (2012) Gendered citizenship for deepening local democracy: A study of ten villages in Maharashtra. In: Palanthurai G (ed) Elected women representatives and empowerment. Concept Publishing Company Pvt. Ltd, New Delhi, pp 79–96

Borde, C., & Malovany-Chevallier, S. (Trans.). (2009). *The second sex; Simone De Beauvoir.* New York: Vintage Books. https://uberty.org/wp-content/uploads/2015/09/1949_simone-de-beauvoir-the-second-sex.pdf. Accessed on 5 Jan 2017.

Census of India. (2011). Retrieved from www.Censusindia.gov.in

Chattopadhyay, R., & Duflo, E. (October 2001/2003, Revised). *Women as policy makers: Evidence from a India-wide randomized policy experiment* (pp. 1–21). weber.ucsd.edu/~tkousser/WomenasPolicyMakers.pdf. Accessed on 20 June 2013.

CMAO (City Managers' Association Orissa) (2009) Local government management learning kit for elected women representatives of urban local bodies of Orissa. Housing and Urban Development Department, Govt. of Orissa, Orissa. www.cmao.nic.in/2009/Events/LGM%20Women%20Orissa%20Module.pdf. Accessed on 26 November 2013

Deepmala K (2012) Women's reservation bill, a triumph or a mirage. In: Palanthurai G (ed) Elected women representatives and empowerment. Concept Publishing Company, New Delhi, p 99

Dikshit RD (2008) Geographical thought a contextual history of ideas. PHI Learning Pvt. Ltd, New Delhi

Doepke M, Miche'le T, Voena A (2012) The economics and politics of women's rights. The Annual Review of Economics 4(6):6.1–6.34. https://doi.org/10.1146/annurev-economics-061109-080201

Drage, J. (2001). *Women in local government in Asia and the Pacific: A comparative analysis of thirteen countries* (Summit of Women Mayor and Councillors: United Nations Economic and Social Commission for Asia and the Pacific). www.ucl.ac.uk/.../UNESCAP_Drage_Women_Local_Government_Asia_Pacific.pdf. Accessed on 09 Mar 2018.

Franchi, L. (n.d.). *Sartre and freedom.* https://www.scribd.com/document/61230001/Sartre-and-Freedom. Accessed on 17 July 2017.

Garrett, J. (2005). *Amartya Sen's ethics of substantial freedom* (pp. 1–7). http://people.wku.edu/jan.garrett/ethics/senethic.htm. Accessed on 1 May 2016.

Ghosh, A. (2003). Women's reservationa and electoral politics in urban local bodies: an analysis of Chennai municipal corporation elections. Indian Journal of Gender Studies 10(1):133. https://doi.org/10.1177/097152150301000106

Ghosh, J. (2009). *Transcending gender reservation, representation and identity among elected women representatives in municipal politics* (Published Master of Science dissertation). School of Interdisciplinary Area Studies, University of Oxford. www.southasia.ox.ac.uk/.../Ghose%2C%20J%20MSc%20thesis%20June%202009.pdf. Accessed on 01 Apr 2013.

Ghosh, A. (2010). Role of women councillors in urban local bodies: A case study of the Kolkata and the Howrah Municipal Corporation. Pragatishil Prakashak, Kolkata

Ghosh A (2011) Elected women representatives of West Bengal: Experience and condition in Birbhum, SWS-SRTT occasional paper 10. Jadavpur University, School of Women's Studies, Kolkata

Ghosh A, Lama-Rewal ST (2005) Democratization in progress, Women and local politics in urban India. Tulika Books, New Delhi

Gupta S (2006) Political empowerment of women in urban local bodies and the 74th constitution amendment: Dispelling doubts and devising future strategies. In: Singh UB (ed) Empowerment of women in urban administration, Experiences and strategies. Serial Publications, New Delhi, pp 67–80

Haug, M. (2012). Women's political participation at the sub-national level in Nepal: A mapping study (UN WOMEN). In *South Asia Country review series on women's political leadership* (pp. 5–20). Norway: Norwegian Institute for Urban and Regional Research; UN Women. https://asiapacific.unwomen.org/en/digital-library/publications/2012/6/womens-political-participation-at-the-subnational-level-in-nepal-a-mapping-study. Accessed on 30 Apr 2013.

Holzner M, Wit JD (2003) Supporting decentralised urban governance: Training women municipal councillors in Mumbai, India, Working paper series no 386 Institute of Social Studies Netherland. Institute of Social Studies Netherland, Rotterdam. scholar.google.com/citations?user=6iT1FRIAAAAJ&hl=en. Accessed on 30 Jan 2014

Horowitz L (2009) Getting good governance for women: A literature review, Agriculture and rural development, discussion paper 43. World Bank, Washington, DC, pp 39–45. siteresources.worldbank.org/EXTARD/.../GenderandGovernanceLitReview.pdf. Accessed on 7 Nov 2012

ICRW (International Centre for Research on Women) (n.d.) Country reviews women's political participation in South Asia, Discussion paper series on women's political leadership. UNWOMEN, New York, pp 1–23. https://iknowpolitics.org/sites/default/files/country-reviews-womens-political-participation-in-south-asia.pdf

Iddo L (2012) Sartre's absolute freedom in being and nothingness: The problems persist. Philosophy Today 56(4):463. https://www.questia.com/library/journal/1P32830614741/sartre-s-absolute-freedom-in-being-and-nothingness. Accessed on 25 July 2017

International Labour Organization. (2017). *World Employment and Social Outlook (WESO): Trends for women 2017*. Geneva: PRODOC, ILO. http://www.ilo.org/wcmsp5/groups/public/%2D%2D-dgreports/%2D%2D-inst/documents/publication/wcms_557245.pdf. Accessed on 22 Jan 2018.

IPU (2017) Women in national parliaments. IPU, Dwarka. http://www.ipu.org/wmn-e/world.htm. Accessed on 29 June 2017

IWRAW, Asia-Pacific. (n.d.). *Facilitating the fulfilment of state obligations towards women's equality baseline report women and political participation in India. 01-120*. Accessed from http://www.iwraw-ap.org

Iyer L, Mani A, Mishra P, Topalova P (2011) The power of political voice: Women's political representation and crime in India, Working paper 11-092. Harvard Business School, Boston, pp 2–34. www.hbs.edu/faculty/Publication%20Files/11-092.pdf. Accessed on 26 Sept 2012

John ME (2012) Women in power? Gender, caste and the politics of local urban governance. In: Raghunandan TR (ed) Decentralisation and local governments the Indian experience essays from economic and political weekly. Orient Blackswan, New Delhi, p 349

Johnston, R. J. (1997, Reprinted 2000). Geography and geographers: Anglo American Human Geography since 1945. Delhi: Replika Press Pvt. Ltd.

Joseph V (2011) Capability of elected women representatives on gender issues in Grassroot governance. Concept Publishing Company Pvt. Ltd., New Delhi

Khattak SG (1996) Women and local government, Working paper series 24 Sustainable Development Policy Institute. Sustainable Development Policy Institute, Islamabad, pp 3–16. https://sdpi.org/publications/files/W24-Women%20and%20Local%20Government.pdf. Accessed on 29 Jan 2013

Kodoth P, Mishra US (2011) Gender equality in local governance in Kerala. Economic and Political Weekly XLVI(38):36–43. https://www.epw.in/journal/2011/38/insight/gender-equality-local-governance-kerala.pdf. Accessed on 02 July 2012

Kulkarni MN (1996) Madhya Pradesh evolving a women's policy. Economic and Political Weekly 31(12):713–715. https://www.epw.in/journal/1996/12/commentary/madhya-pradesh-evolving-women-s-policy.pdf
Kumar NA (2006) Do women have efficiency to run the urban administration? In: Singh UB (ed) Empowerment of women in urban administration, Experiences and strategies. Serials Publications, New Delhi, pp 105–114
Mandal A (1999) Political empowerment of women at the grassroots [review of the book *Political Empowerment of Women at the Grassroots* by K. C. Vidya]. Indian Journal of Gender Studies 6(2):333. http://journals.sagepub.com/doi/abs/10.1177/097152159900600213. Accessed on 11 Mar 2013
Mazumder V (1989) Reservations for women. Economic and Political Weekly 24(50):2795–2796
Mishra P (2015) Women's reservation bill: A long awaited dream. In: Sarkar DS, Pandey DP (eds) Women in politics in South Asia. Manglam Publications, Delhi, pp 62–63
Morgan S (2003) Equality and diversity in local government in England, a literature review. Local and Regional Government Research Unit, ODPM, London. http://www.local.odpm.gov.uk/research/crosscut/equality/equal.pdf. Accessed on 7 Nov 2012
Nair J (2012) All in the family? Gender, caste and politics in an Indian Metropolis. In: Banerjee N, Sen S, Dhawan N (eds) Mapping the field, Gender relations in contemporary India, vol 2. Mandira Sen for STREE, Kolkata, pp 464–497
Nakkeeran K (2012) EWRs' representation, responsiveness and accountability in grassroots governance. In: Palanthurai G (ed) Elected women representatives and empowerment. Concept Publishing Company Pvt. Ltd., New Delhi, pp 59–78
Narain K, Mridubhashini M (2006) Dynamics of empowering women in urban development. In: Singh UB (ed) Empowerment of women in Urban Administration, Experiences and strategies. Serial Publications, New Delhi, pp 184–193
Nayak GC (1998) The concept of freedom in Sartre and Sankara. Indian Philosophical Quarterly XXV(1):119–120
Omvedt G (2005) Women in governance in South Asia. Economic and Political Weekly 40(44–45):4746–4752
Palanthurai G (2012) Elected women representatives and empowerment. Concept Publishing Company Pvt. Ltd, New Delhi
Palanthurai G, Uma G, Vanishree J (2009) Networking of elected women representatives at grassroots. Concept Publishing Company, New Delhi
Panday PK (2008) Representation without participation: Quotas for women in Bangladesh. International Political Science Review 29(4):489–512
Paradza G, Mokwena LG, Richards R (2010) Assessing the role of councillors in service delivery at local government level in South Africa (Rep. No. 125). Centre for Policy Studies, Johannesburg. www.cps.org.za. Accessed on 28 May 2013
Peet R (2004) Modern geographical thought. Blackwell, Oxford
Phillips A (2003) The politics of presence. Oxford University Press, Oxford. ISBN 9780198294153
Rai P (2011) Electoral participation of women in India (Key Determinants and Barriers). Economic and Political Weekly XLVI(3):47–55
Raj MR (2012) Empowerment of women elected members of Kerala: What next? In: Palanthurai G (ed) Elected women representatives and empowerment. Concept Publishing Company, New Delhi, pp 167–179
Raju S, Lahiri-Dutt K (eds) (2011) Doing gender doing geography; emerging research in India. Routledge, New Delhi
Ranjan IS, Retnakumar J (2005) Women's reservation bill: Some emerging issues. Economic and Political Weekly 40(39):4191
Rao, B. (2019). *Women MPs in Lok Sabha: How have the numbers changed?* https://factly.in/women-mps-in-lok-sabha-how-have-the-numbers-changed/. Accessed on 16 May 2019.
Rizvi SH (2006) Empowering women through reservation in urban local bodies-difference between theory and practice. In: Singh UB (ed) Empowerment of women in urban administration, Experiences and strategies. Serials Publications, New Delhi, pp 193–211

Ruet J, Lama-Rewal S (2005) Governing India's metropolises. Routledge Publication, New Delhi
Rukmini, S. (2017). *More women in India's workforce would make India much richer, says ILO*. http://www.huffingtonpost.in/2017/06/15/more-women-in-indias-workforce-would-make-india-much-richer-sa_a_22250501/. Accessed on 22 Jan 2018.
Ryan AB (2006) Post-positivist approaches to research. In: Researching and writing your thesis: A guide for postgraduate students. MACE: Maynooth Adult and Community Education, Maynooth, pp 12–26. eprints.maynoothuniversity.ie/874/. Accessed on 7 Apr 2018
Sarkar, D. S. (2015). Politics and women rights in India. In D. S. Sarkar & D. P. Pandey (Eds.), *Women politics in South Asia* (pp. 186–187; 194–195). Delhi: Manglam Publications.
Sathe D, Klasen S, Priebe J, Biniwale M (2013) Can the female Sarpanch deliver? Economic and Political Weekly XLVIII(11):50–57
Sathiabama K (2012) A critique on the political reservation and its utilization by the elected women representatives. In: Palanthurai G (ed) Elected women representatives and empowerment. Concept Publishing Company, New Delhi, p 197
Sekhson J (2006) Engendering Grassroot democracy: Research, training and networking for women in local self-governance in India. NWSA Journal, JSTOR 18(2):101–122. http://www.jstor.org/stable/4317209. Accessed on 19 June 2013
Sen A (1993) Capability and well being. In: Nussbaum M, Sen A (eds) The quality of life. Oxford University Press, New Delhi, pp 9–29
Sen A (1999) Inequality reexamined. Oxford University Press, New Delhi
Sen A (2000) Development as freedom. Oxford University Press, New Delhi
Sheel R (2003) Women in politics in Japan. Economic and Political Weekly 38:4097–4101. http://www.epw.in/journal/2003/39/commentary/women-politics-japan.html. Accessed on 19 Dec 2012
Singaray P (2014) West Bengal's election story the caste question (Commentry). Economic and Political Weekly XLIX(17):10–12
Singh SK (2006) Women empowerment in India: Emerging leadership in urban U.P. In: Singh UB (ed) Empowerment of women in urban administration, Experiences and strategies. Serials Publications, New Delhi, pp 61–64
Singla P (2011) Women's participation in Panchayati Raj nature and effectiveness. Rawat Publications, New Delhi
Srivastava DN (2015) Political status of Indian women: Progress since independence. In: Sarkar S, Pandey PK (eds) Women in politics in South Asia. Mangalam Publications, New Delhi, pp 199–213
Stewart K (2006) Designing good urban governance indicators: The importance of citizen participation and its evaluation in Greater Vancouver. Cities 23(3):196–204
Sundstrom, A. (2008). Leaving the 'proxy woman or politician' dichotomy: A qualitative study of the possibilities and obstacles for elected women's participation in Indian local governance (BA Thesis in Political Science, *Minor Field Study*). https://gupea.ub.gu.se/bitstream/2077/17844/1/gupea_2077_17844_1.pdf. Accessed on 20 June 2020.
Upadhyay KK, Bhardwaj P (2006) The gender paradigm in local governance: Indian experience. In: Singh UB (ed) Empowerment of women in urban administration, Experiences and strategies. Serials Publications, New Delhi, pp 130–148
West Bengal State Election Commission. (2010–2013). Accessed from www.wbsec.gov.in
World Bank. (2011). *World development report 2011 conflict, security and development*. https://openknowledge.worldbank.org/handle/10986/4389
Yupanqui, A.M. (2011). *Amartya Sen's notion of freedom a conceptual history of a universalist presumption* (SIMT29, Master's thesis in developmental studies). Faculty of Social Science, Graduate School, Department of Political Science, Lund University: Spring. lup.lub.lu.se/student-papers/record/1966976/file/1976430.pdf. Accessed on 2 June 2016.

Chapter 2
Urban Local Governance in West Bengal

> "It is necessary that the people and their representatives are fully involved in the planning and implementation of the programmes at local level. If democracy in Parliament and State Legislatures is to remain strong and stable, its roots must reach towns and villages and the cities where the people live." (MoHUA n.d.)

2.1 The 74th Constitutional Amendment Act and Its Background

The Government of India declares the above statement in the background paper for the 74th Constitutional Amendment Act (CAA), which clarifies clearly why India headed for urban local governance reforms under this CAA (MoHUA n.d.). When the Indian constitution was written, there was no clear-cut directive on the structure of local government. But during the planned development period of India, it was felt that to develop the cities as regional growth centres to cater to their respective regions, there is a need for proper planning and management at the local level. Therefore, the decentralization of urban governance process was only initiated by the 74th CAA in India in 1992. Before that, all the urban areas were controlled by the state government or their assigned agencies, which usually made the cities weaker and dependent on the state governments. Thus, through the 74th CAA, the urban local governments got linked to the national government structure, and as Rao (2016: 26) has pointed out, it was meant to develop the government closest to the citizens to involve them in the decision-making process, to improve their living conditions as well as to make use of their knowledge and capabilities in achieving all-round development.

To clarify the proposed objectives linked to enhancement of urban local governance, the 74th Constitutional Amendment Act introduced a new part called Part IX A, which mentions the specific provisions under this act. These provisions include

R. Banerjee, G. Samanta, *Negotiating Terrain in Local Governance*, Local and Urban Governance, https://doi.org/10.1007/978-3-030-60663-3_2

the following: (a) the constitution of urban local government or municipalities will follow the modalities and norms developed by their respective state governments; (b) the urban local government or municipalities will be constituted and formed by the elected representatives from each and every ward of a city; (c) to ensure the participation of the citizens in the planning and management of the city, a ward committee for each and every ward of the cities having more than 3 lakhs population should be constituted; (d) the tenure of local urban governments would be 5 years, and regular election should be carried out at a regular interval of 5 years under the direct control of the State Election Commission; (e) municipalities as independent local government should perform certain functions which are listed in the Twelfth Schedule of the Constitution (Table 2.1); (f) the financial provision of the local government on revenue collection and expenditure will be controlled by the respective state's financial provision and laws; (g) the State Finance Commission will monitor the financial condition of urban local governments and will make recommendations accordingly; (h) the maintenance of accounts of the municipalities and their regular auditing will be conducted following the state financial rules; and (i) for decentralized and effective planning of urban areas, district planning committee for smaller cities and metropolitan planning committee for the cities of more 1 million population would be constituted.

In addition to the other provisions for regular maintenance, development and planning of municipalities, the 74th Constitutional Amendment Act gave a strong priority to reservation of seats for scheduled castes (SC), scheduled tribe (ST) population and women. The document (MoHUA n.d.) clearly states:

Table 2.1 XII schedule (essential functions of urban local bodies)

1	Urban planning including town planning
2	Regulation of land use and construction of buildings
3	Planning for economic and social development
4	Roads and bridges
5	Water supply for domestic, industrial and commercial purposes
6	Public health, sanitation conservancy and solid waste management
7	Fire services
8	Urban forestry, protection of the environment and promotion of ecological aspects
9	Safeguarding the interests of weaker sections of society, including the handicapped and mentally retarded
10	Slum improvement and upgradation
11	Urban poverty alleviation
12	Provision of urban amenities and facilities such as parks, gardens, playgrounds
13	Promotion of cultural, educational and aesthetic aspects
14	Burials and burial grounds; cremations, cremation grounds and electric crematoriums
15	Cattle pounds; prevention of cruelty to animals
16	Vital statistics including registration of births and deaths
17	Public amenities including street lighting, parking lots, bus stops and public conveniences
18	Regulation of slaughter houses and tanneries

In order to provide for adequate representation of SC/ST and of women in the municipal bodies, provisions have been made for reservation of seats. The proportion of seats to be reserved for SC/ST to the total number of seats shall be same as the proportion of the population of SC/ST in the municipal area. The reservation would be made in respect of seats to be filled by direct elections only. Not less than one-third of the total number of seats reserved for SC/ST shall be reserved for women belonging to SC/ST. This is a mandatory provision. In respect of women, the seats shall be reserved to the extent of not less than one-third of the total number of seats. This includes seats reserved for women belonging to SC/ST.

2.2 Changing Concept of Governance/New Governance[1]

The meaning and the issues of governance have undergone significant changes in the last two decades along with the process of globalization. The structural adjustment programmes and the realization that the government sector cannot provide urban services in the cities over a long run on a sustainable basis have made the cities more dependent on external funding, as well as on new partnerships with other agencies such as the private sector, non-government organizations (NGOs) and community-based organizations (CBOs). To accommodate these changes in running cities, and in making provision for the fast increasing demand of urban basic services, the governance process has become a serious matter of concern in developing countries like India (Samanta 2013).

Over the past two decades, the process of globalization has become all-pervasive in India with increasing impact on our everyday life. The urban sections are much more influenced by globalization than any other individual sector in India (Sridharan 2008). However, Harris (2003) opines that the twin processes of globalization and decentralization are not the sole factors working in cities in this period. Several processes are refashioning cities simultaneously, among which a few are the cities' continual change and incremental adjustment to external markets, long-term process of deindustrialization, impact of increased world competition and so on. Liberalization and macroeconomic reform has led to the privatization and reorganization of urban basic services. As a result of this, there has been a tremendous increase in the number of agents providing public sector services. This multiplicity of private agencies; government agencies from national, provincial and local levels; NGOs; CBOs; and charitable institutions has made service delivery a complex issue (Samanta 2010a).

Governance does not merely refer to the exercise of governmental authority, rather is an exception of the notion of government to include people into the system

[1] Part of this chapter was earlier published in Samanta, G. (2020). Decentralized governance versus state dependence: Financial challenges and participatory development in small cities of West Bengal. In R.R. Thakur, A. K. Dutt, S. K. Thakur and G. M. Pomeroy (eds), *Urban and Regional Planning and Development: 20th Century Forms and 21st Century Transformations* (pp. 321–336). Springer.

in the forms of collective decision-making, participatory and representative system at all levels. Urban governance, therefore, can be described as the means through which inhabitants and the political and bureaucratic apparatus of government institutions in any urban area reach an agreement on how to progress towards the achievement of multiple goals and how best to use limited public resources and capacities (Roy 2006).

This changing concept of governance is defined as 'new governance'. The international development assistance to the cities of developing countries has started to impose conditions of changing earlier urban governance processes into a new responsive and good governance. The United Nations Development Programme policy document argues that good governance and human development are indivisible and the former is a necessary ingredient to achieve equitable and sustainable development (Datta 2006). In this new governance process, a lot of emphasis has been given to a number of aspects of governance such as decentralization, people's participation, accountability, transparency and so on. To improve urban infrastructure and to provide urban services effectively to the poor, urban governance reform is considered as an urgent necessity in JNNURM, the latest programme for urban areas in India. Restructuring the system of governance and reduction of public expenditure on infrastructure and civic amenities has, thus, been perceived as the most crucial element of the strategy to meet the challenge of the emerging crisis in the provision of urban basic services (Kundu 2003).

2.3 Decentralized Governance and Participatory Development in West Bengal

The decades following the Independence of India experienced emphasis on planning in every sector. The urban sector was no exception. The planning regime in urban development continued for decades with overwhelming importance being placed on large and metropolitan cities in India, through the Master Plan, the Town and Country Planning Act, etc. In the 1980s, there was a change in the emphasis of planned urbanization following the recommendations of the National Commission on Urbanization (NCU). The report of the NCU published in 1988 prioritized the development of small and medium towns, which together constitute a major share of urban population in India. The early 1990s also experienced another significant change in the management procedures of cities, prioritizing the decentralized model of governance instead of the centralized model of 'planning'. The 74th Constitutional Amendment Act, passed in 1992, tried to ensure these changes. The people-centric model of governance became an important tool to improve the quality of life of the urban people through the means of enhanced urban basic services.

Besides analyzing the background of the 74th Constitutional Amendment Act and the change in the concept of governance for better servicing the cities, this chapter traces the changes in the process of urban governance in West Bengal, with a special focus on small cities. Decentralization and people's participation have become important components of urban governance in West Bengal since the 1980s. The municipal financing system has become more transparent. This chapter looks into the state-level changes in the urban governance structure in West Bengal and their impacts on the efficiency of the provision of basic services in smaller cities. The governance process and its changes have been understood through a rigorous reading of government documents from both the state and the local governments and a series of interviews with government officials and elected representatives at both levels. The impact of the policies on the citizens has been studied through in-depth empirical research, following methods starting from detailed questionnaires to informal individual interviews and group discussions with the citizens at the household level. The common element among the selected cities is their location away from the metropolitan city of Kolkata, as they all belong to the secondary city group when it comes to their relative importance in the state.

2.3.1 Reforms in Urban Governance in West Bengal

Restructuring and reforms in urban governance had been initiated in West Bengal much before the 74th Constitutional Amendment. Among the newer aspects of governance such as decentralization, people's participation, accountability and so on, decentralization is an important one, and in West Bengal, decentralization took place much faster than in the other Indian states (World Bank 2006). This process of decentralization has made the cities in West Bengal self-governed local bodies, and has made people's participation mandatory in the process of urban governance. Institutional autonomies are enjoyed by the city governments here, although their financial dependence on the state government is still very high. This financial dependence is higher in smaller cities, and therefore, the efficiency of the city governments in the provision of basic services depends mostly on financial grants from the state government. When the financial condition of the state is not good, there is a cut in the state government grants, which ultimately impacts the functioning of the small cities. The number of statutory towns or urban local bodies (ULB) and the population living in existing cities are both increasing. The increase of population in urban areas is occurring at a much faster rate than the capacity enhancement of resource generation in these cities. Moreover, the financial condition of the state is not good enough to bear the cost of the increasing urban area and population. The impact of all these together is going to be high, affecting the basic services in the cities.

Around 31.87% of the state's total population lives in 127 statutory ULBs and 780 census towns[2] of West Bengal, occupying hardly 2.93% of the total land area. The high rate of increase in urban population (Table 2.2) is not compatible with the

[2] The settlements attaining urban characteristics such as density of population and non-farm activity, but not notified or administered as statutory town.

Table 2.2 Urban West Bengal, 1961–2011

Year	Percentage of urban population	Percentage growth of urban population	Number of census towns	Number of statutory towns	Percentage of urban population in Kolkata Metropolitan Area
1961	24.40	35.97	–	80	–
1971	24.70	28.41	48	83	64.11
1981	26.50	31.73	89	87	63.64
1991	27.48	29.49	148	106	58.92
2001	28.03	20.20	252	123	58.88
2011	31.89	–	780	127	48.44

Source: ILGUS 2002 and 2011 Census

growth in the cities' economy. As a result, among the total urban population of the state, about 40% live in slums (Mitra and Mukherjee 2008). The state has not performed well on the economic front for the last couple of decades. In terms of per capita state domestic product (SDP), West Bengal declined from being first in 1960 to being the fifth among Indian states by 2014–2015. The share of the state in the total industrial production of the country had been reduced from 24% in 1947 to 11.9% in the late 1970s (Khasnabis 2008) and to 4.38% in 2012–2013. In terms of human development, the state's position has changed positively from 0.492 in 2007–2008 to 0.604 in 2015. However in terms of rank among Indian states, its position is still tenth.

One significant characteristic of the recent urbanization in West Bengal is that the percentage share of urban population outside the metropolitan area has increased considerably (Table 2.2), which indicates that more urban areas are coming up outside the metropolitan area and its shadow zone. Although the rate of urbanization has increased significantly, the notification of statutory towns since 2001 has been very limited in West Bengal (Table 2.2). On the contrary, census towns have increased considerably over the last two decades. Although there are several debates regarding the role of census towns in measuring the level of urbanization (Kundu 2011; Pradhan 2013), there are many census towns whose agglomerated urban population exceeds the minimum population requirement of 30,000 for being a statutory town in the state (Samanta 2014, 2017). The process of municipalization has slowed down significantly in the last two decades. This situation may be linked to the bad financial condition of the state of West Bengal, as the cost of developing and maintaining a statutory town is a state affair.

The governance structure of ULBs in West Bengal is a decentralized one. The chief characteristic features of the municipal government are well narrated in the GOWB Report (2003). The ULBs are divided into smaller areas called wards. Each ward is represented by an elected member, a councillor. One-third of the total seats of a municipality are reserved for women. Seats are proportionately reserved for the weaker sections (Scheduled Caste and Scheduled Tribe) as well. Municipal elections are governed at a regular interval of 5 years under the direct supervision of the State Election Commission. Mayors/Chairpersons of the municipal bodies should be elected by the councillors. The municipal corporations (big cities) and municipal towns are governed by a cabinet formed of the local self-government through the Mayor-in-Council/Chairman-in-Council model, respectively. Respective municipalities appoint their officers and employees with approval of the state government. There should be Municipal Accounts Committees in the ULBs to keep vigil on municipal accounts. Municipal bodies are responsible for urban planning, socio-economic development of its residents, especially the underprivileged ones, and for providing infrastructure for basic civil services. Constitution of ward committees in each municipal ward is mandatory. To ensure transparency and accountability in municipal administration, each ward committee is required to hold at least two general meetings annually, where the citizens can participate in the development plans and programmes.

2.3.2 Pre-74th Constitutional Amendment Act Period

The state of West Bengal started the decentralized planning experiment in 1978, following the recommendations of the Urban Development Strategy Committee. The major recommendations of the committee (Datta 1990) were as follows:

- The difference in per capita investment between ULBs in Calcutta Metropolitan Area (now Kolkata Metropolitan Area), and those in non-Calcutta Metropolitan Area should be brought down by tilting the balance in favour of the latter, in order to restrict migration to the metropolis and to develop the latter as growth centres.
- The difference between development grants and per capita development expenditure should be reduced.
- The use of local resources and materials should be, as far as possible, maximized.
- Special efforts need to be made for improving the quality of life of the urban poor and the weaker sections.
- The ULBs should be entrusted with the responsibility of preparation of development plans and implementation of development programmes in a decentralized manner.

Following these recommendations, the West Bengal government practiced many facets of the 74th Constitutional Amendment Act, before it got passed by the parliament, such as decentralization of development programmes, especially the preparation and implementation of development programmes directly by the ULBs. A major step in the direction of change in governance in West Bengal started in 1978, with the formation of two independent departments in the state government for urban areas, namely, Municipal Affairs and Urban Development. Both the departments function under a single ministry of Municipal Affairs and Urban Development. In the 1980s and the early 1990s, a number of new institutions such as the Directorate of Local Bodies (DLB), Municipal Engineering Directorate (MED), Institute of Local Government and Urban Studies (ILGUS), State Urban Development Agency (SUDA) and West Bengal Valuation Board (WBVB) were created as support agencies to give administrative and technical assistance for capacity building of the ULBs to accommodate change in governance more efficiently. However, as part of a recent administrative reform of merging different departments of the state government, the Municipal Affairs and the Urban Development Departments have been merged in 2017 for better coordination of the two departments in charge of urban development, administration and basic services (GOWB 2017). Among the other reform initiatives taken up by the state government, a few important ones are the following:

- Minimizing the gap between income and expenditure of the urban local bodies
- Enhancing the bases of own revenue collection.
- Giving priority to the basic services for the urban poor and weaker sections of the society.
- Decentralization of the development projects.
- Making the municipalities capable to provide urban services as well as micro-planning for their area.
- It is made mandatory for the ULBs to prepare Draft Development Plans (DDP) covering a period of 5 years for planned development.
- The MAD has made outlines for the Health Care Service Delivery Improvement Plan and Education Improvement Plan of the Draft Development Plan which can be followed by the ULBs to make their health and education plans in advance.

Besides empowering the ULBs, another important recommendation was to reduce the gap in the per capita investment for infrastructure and services between the ULBs of Kolkata Metropolitan Area (KMA) and ULBs of non-KMA area. This is a crucial aspect of urban governance and planning in West Bengal. There are clear-cut divisions between these towns and cities in terms of monetary allocation under different heads. The towns located in the KMA area get more grants, more programmes and facilities from the state government. Although a recommendation was made in 1978 to reduce this gap, it still exists on many fronts.

2.3.3 Post-74th Constitutional Amendment Act Period

Municipal reforms and changes in urban governance have been prioritized in India since the 74th Constitutional Amendment in 1992. This amendment enables urban local bodies to prepare plans for economic development and social justice, to levy taxes and to emerge as forums for the political representation of hitherto marginalized sections of society (Roy 2006). The major thrust among all the recommendations of the 74th Constitutional Amendment Act was on good governance, to make ULBs independent units of self-governance through decentralization and devolution of power and functions and active participation of communities in the governance process. Besides decentralization and participation, good governance also emphasizes upon transparency and accountability.

The West Bengal Municipal Act (1993) listed 49 obligatory functions and 40 discretionary functions for making the ULBs effective local self-government in the state. In this new Act, many services such as health, sanitation and environmental management were defined as obligatory functions of the ULBs, which were not mandatory functions of the ULBs before. Datta (2003) has attributed the achievement of West Bengal in decentralization and participatory governance to the character of the regime rooted in the ideological commitment of the Left Front partners, and the strong political will to strengthen the elected local government. Following many reform initiatives for structural, organizational and institutional changes, the municipal governance process in West Bengal has changed significantly (GOWB 2003). The present characteristics of municipal governance in West Bengal can be outlined as follows (GOWB 2003):

- Municipal areas are divided into several wards mainly depending on the population size of the municipality.
- Each ward is represented by an elected member.
- One-third of the total seats of a municipality are reserved for women.
- Seats are proportionately reserved for the weaker sections (Scheduled Caste and Scheduled Tribe).
- Municipal elections are governed at a regular interval of 5 years under the direct supervision of the State Election Commission.
- Mayors/Chairpersons of the municipal bodies are directly elected by the councillors.
- The corporations and municipal towns are governed by a cabinet formed of local self-government through the Mayor-in-Council/Chairman-in-Council model in the ULBs.
- Respective municipalities appoint their officers and employees with approval of the state government.
- It is mandatory to form Municipal Accounts Committees in the ULBs to keep vigil on municipal accounts. For better and transparent functioning of the urban local bodies, the Municipal Accounts Committee would be headed by the opposition leader. Municipalities are to prepare their respective 5-year Draft Development Plan (DDP) and annual development plan in consultation with the respective ward committees, against which the grant is sanctioned by the state government for the development activities in the cities.
- Municipal bodies are responsible for urban planning, socio-economic development of its residents, especially the underprivileged ones, and providing infrastructure for basic civil services.
- Constitution of ward committees in each municipal ward has been mandatory. To ensure transparency and accountability in municipal administration, each ward committee is required to hold at least two general meetings annually where the citizens can participate in the development plans and programmes.
- The concerned ward committees are to form beneficiary committees for implementation of development programmes and to maintain transparency.

- Municipalities are to prepare their respective 5-year Draft Development Plan (DDP) and annual development plan in consultation with the respective ward committees and the community development societies set up with the women representatives of economically weaker sections.

The 74th Constitutional Amendment Act (CAA) recommended the creation of separate planning committees for districts and metropolitan areas, for finalizing the municipal plans in their ambit, along with the commitment of the state and central assistance for the purpose. The committees are empowered to undertake planning for matters of common interest between the panchayats[3] and the municipalities, and these common interests are physical and natural resources, infrastructure development and environmental conservation (Datta 1995). Immediately following the 74th CAA, the government of West Bengal enacted both the District Planning Committee (DPC) Act and the Metropolitan Planning Committee Act for effective decentralization of the planning process. The West Bengal District Planning Act was passed in 1994, and most of the District Planning Committees in the state have been effectively performing their roles since then. This committee prepares Draft Development Plans for the entire district. Under the District Planning Committee, a sub-committee called the Urban Development Sub-Committee (UDSC) co-ordinates plans and programmes of the ULBs under the DPC's Draft Development Plan (Datta 2003).

The field-level observations from different cities of West Bengal scattered over different districts of the state revealed that the District Planning Committees have effectively monitored and sanctioned both the DDPs and the double-entry accounting system of the ULBs (Samanta 2015). The structures of both the DPC and the UDPC are decentralized, consisting of both district-level and city-level representatives. These committees have been functioning effectively without any interruption since their formation.

2.4 Decentralized Structure of City Governments

West Bengal is the first state in India to accept and operationalize the cabinet form of city governance through the Mayor/Chairman-in-Council model in all the ULBs (Datta 2003). This model, based on the concept of democratic decentralization and collective responsibility, makes the urban governments more accountable. The Mayor/Chairman-in-Council system was introduced in West Bengal in 1984, in which a Mayor or Chairperson, as elected representative, is in a more powerful position than the Municipal Commissioner or Executive Officer, who are administrative heads of the government. Under this system, there is a five to ten members' board of councillors who run the local government with the help of the Mayor or the Chairperson. 'The mayor or chairperson can choose his Council, just like a minister

[3] Lowest unit of local government for rural areas in India, usually covering several villages.

can choose his Cabinet and s(he) allocates the functions among the members of the Council too' (Hamid 2004).

Under the Mayor/Chairman-in-Council, there are several different committees such as the Finance and Resource Mobilization Standing Committee; Solid Waste Management Standing Committee; Water Supply Standing Committee; Public Works Standing Committee; Health Education and Urban Poverty Alleviation Standing Committee; Public Health and Sanitation Standing Committee; Minority, Scheduled Caste and Scheduled Tribe Development Sub-Committee; Municipal Accounts Committee; Mutation and Separation Sub-Committee; etc. However, the number of such committees and their respective names are found to vary widely depending on the respective importance of the activities in the cities. These municipal-level committees are composed of elected representatives and officials from the local government. These are headed by the President and Vice President who are elected representatives. Below the level of municipal committees, there are ward committees in each and every ward of the cities. These ward committees are headed by the ward councillor as well as by the president and the secretary nominated by the ward committee members.

In most states, the state-appointed Municipal Commissioner is the executive functionary, and the executive powers are exercised by a triumvirate set of authorities—the council, the standing committees and the appointed commissioner (Datta 1995). However, in the Mayor-in-Council pattern of the city governance in West Bengal, the Mayor/Chairperson is the designated executive and therefore, exercises greater control and supervision over the Chief Executive Officer, who represents the state bureaucracy in the ULBs. As a result, there is no coercion between the bureaucrats and the elected representatives, as the overall power is enjoyed by the Mayor-in-Council. While on one hand this has helped in decentralization of power amidst the people's representatives and in smooth running of the urban governments, on the other hand, the ruling parties with their overwhelming majority in the city governments have made the governing process far more influenced by political leadership through its elected representatives.

Although political stability in governance helps in pursuing policies and programmes on a continued basis, at least for certain period of time (Ghosh and Ahmad 1996), a very long period of stability sometimes corrupts the system. Because of this, one section of the citizens benefit at the cost of the others due to their similar political identity with that of the city government. This has happened in most of the cities in West Bengal. Lack of sizeable representatives from the opposition in the ULBs has made the Mayor-in-Council more powerful and has enabled them to control the city governance to a large extent. As part of the reforms process for better running of urban administration, a number of new posts have been created in the ULBs. Executive officers have been recruited in each and every ULB in the recent past.

2.5 Institutional Reforms and Basic Services

Improvement of basic services and the necessary institutional reforms in the cities of West Bengal started with external funds. Kolkata Urban Services for the Poor (KUSP) was an externally funded programme of the government of West Bengal to improve the urban basic services and quality of life, especially for the poor people living in the urban areas. The programme was initiated through a tripartite agreement between the Government of India, the Government of West Bengal and the Department for Urban Development (DFID) of the UK in 2003, and continued up to 2012. This programme seeks to transform the urban poor from mere bystanders in the development process to involved participants of it, and to increase the institutional capacity of the ULBs (Basu 2006). In the first phase, it was implemented in 40 ULBs of the Kolkata Metropolitan Area (KMA). In the second phase, it was introduced in the rest of the ULBs of West Bengal, located outside the KMA. The basic objectives of this programme, as stated by Biswas (2003) were as follows:

I. To improve urban planning and governance in the ULBs by supporting the strengthening of ULBs in line with the 74th Constitutional Amendment and transforming support agencies at the state level, such as Municipal Affairs Department (MAD), Municipal Engineering Directorate (MED), Directorate of Local Bodies (DLB), Institute of Local Government and Urban Studies (ILGUS), Kolkata Metropolitan Development Authority (KMDA) and State Urban Development Agency (SUDA) to enable them to better serve the needs of the ULBs
II. Improved access of poor people to urban services in the ULBs by engaging the slum communities in the provision of basic services through needs identification, prioritization and community contracting
III. Economic development by promoting livelihood opportunities through participatory approaches

The KUSP Programme was implemented by the Municipal Affairs Department (MAD), Government of West Bengal, through the support agencies to strengthen the capacity of the ULBs. It included a number of key players and stakeholders at different levels of operation. The main inventory feature of the KUSP was to focus on the capacity building of the ULBs in order to deliver proper services based on participatory 'bottom-up' planning. By putting the main thrust on prioritization and affordability in the preparation of the Municipal Draft Development Plans (DDP), the KUSP established a demand-side approach and demand management practices in the basic services provided to urban citizens in the cities of West Bengal (Samanta 2010b).

For the effective implementation of the KUSP Programme, a Change Management Unit (CMU) was created to work within MAD. The CMU was headed by a project director with the support of the project manager. The unit had a range of specialists either recruited or contracted as consultants. The CMU comprised of experts from different institutions of the state government such as Municipal Finance, Organization, Development, Urban Planning, Poverty Monitoring, Health, Information Technology and so on. The CMU played both the roles of the programme manager and facilitators of change, working with the support agencies and the ULBs (Biswas 2003). After the KUSP Programme ended in 2012, the provision of basic services became the responsibility of the ULBs with financial support from the state government.

Following the Town and Country Planning Act of 1979, the state of West Bengal has nine development authorities for planning of big urban centres and their surrounding areas. These authorities are to facilitate industrial, infrastructure and social development activities in their demarcated areas. They also perform statutory functions like preparation of Land Use Development and Control Plan (LUDCP) and are responsible for identifying growth centres having economic potential, and for providing support in attracting and facilitating investment in the industrial sector of the state (GOWB 2003). For example, Burdwan town comes under the Burdwan Development Authority (BDA) established in 2002, and Durgapur comes under the Asansol Durgapur Development Authority (ADDA) which was established in 1980. These development authorities are under direct regulation of the Urban Development Department of the state government.

The cities coming under these statutory planning bodies usually get some infrastructural support from the development authorities. These authorities usually control the physical infrastructure of the cities and their newly developed areas. In different cities, the development of city centres, shopping malls, parks and industrial estates and the development of housing estates have been observed under the infrastructural development of the development authorities. Recent trends show that most of the development efforts done by these planning bodies are made under public–private partnerships. Although the urban basic services are completely under the domain of the ULBs, the management of solid waste and other infrastructures sometimes receives support from these planning organizations, that is, the development authorities.

2.6 New Governance Process

The new governance process puts more emphasis on the devolution of power to the people and, therefore, demands accountability, transparency and participation of the community in the decision-making processes of ULBs. In West Bengal, several initiatives have been taken to ensure accountability, transparency and the participation of communities in the urban local governance as part of the urban governance reforms.

2.6.1 Accountability

Among the different aspects of the new governance, accountability has a special role in improving the service delivery mechanism. The general weakness of accountability mechanisms is considered a major barrier to the improvement of basic services. Clearly, strengthening accountability mechanisms and sparking greater citizen demand for change are important pillars of any strategy aimed at improving services in India (World Bank 2006). According to the United Nations Millennium Project (2005), accountability is needed to impel individuals and institutions in the right direction. Accountability mechanisms are essential to hold the government and other service providers responsible for their action (or inaction) in the improvement of services.

Liberalization has allowed market forces to operate in Indian cities to produce basic urban services, and therefore, market accountability needs to be ensured for the citizens. In the competitive urban market, the service providers are obliged to become more accountable to retain and expand their businesses. In Indian cities, the ULB is the prime service provider. However, as a service provider, they also have to be accountable to the citizens. Improving access to information (including prices, forms, officers to be contacted, grievance redressal and procedures to be negotiated by applicants) has been identified as a process to make the local governments more accountable. Publication of the citizens' charter with all the detailed information on the service provision and establishing a grievance redress cell to receive complaints from the citizens are essential tools for ensuring accountability.

According to Datta (1995), incorporating the citizens' charter in the municipal legislation makes the municipal services more responsive and accountable to their clients, thus improving the quality of service and involving the people's choices. Publishing and circulating a citizen's charter have become mandatory in the ULBs of West Bengal to make them more accountable as part of the reforms process under KUSP. All the case studies show that the ULBs have published their citizens' charter and upgraded those from time to time to make the urban government more responsive and more accountable.

In addition to the citizens' charter, hoardings displaying information about municipal services are also observed in different nodal points of the four cities under study. However, the household-level survey results show that the number of citizens lodging complaints against the municipality is still very low, and even for that minority, not much action has been taken by the municipality yet. Therefore, it is clear that although institutional reforms have taken place, it is yet to make the municipalities more accountable in service delivery.

2.6.2 Transparency

Transparency in the ULBs' activities, especially in finance and decision-making, is an important prerequisite of good governance. As part of the new reforms process to ensure transparency in financial dealings, a Municipal Accounts Committee is formed in most of the ULBs in West Bengal. This committee, chaired by the opposition leader, has the power to monitor and supervise the municipal accounts as a whole. However, our field observation revealed that in most cities, the Municipal Accounts Committee could not perform its role due to negligible representation of the opposition in the local government. The ULBs in West Bengal are in most cases controlled by the ruling party of the state, and therefore, even if the opposition party members exist in the municipality, the Municipal Accounts Committee does not perform at all (Samanta 2015).

As part of the municipal reforms, the accounting system of the cities in West Bengal has been transformed from single-entry accounting to double-entry accrual-based accounting. An amendment to the West Bengal Municipal (Financial and Accounting) Rules (1999) has been made in order to provide necessary guidelines on how to do this. Single-entry cash basis of accounting was used to record transactions only on receipt and payment of cash (for income and expenditure items). As a result, an analysis of the true and fair view of the activities of the ULBs was not possible as income accrued (but not received) and expenditure incurred (but not paid for) were not reflected in the financial statements of the ULBs. The new accounting system follows (a) accounting standards issued by the Institute of Chartered Accountants of India; (b) West Bengal Municipal (Finance and Accounting) Rules (1999); and (c) the National Municipal Accounts Manual issued by the Ministry of Urban Development, Government of India (Samanta 2020).

This new system has been introduced in the municipalities of Kolkata Metropolitan Area (KMA) from the financial year 2006–2007 and later on in the ULBs outside KMA. By the end of the KUSP Programme in 2012, all the municipalities had transformed their accounting system from single entry to double entry. This reform in municipal accounting can be considered as a step forward in making the urban governance more transparent.

2.6.3 *Participation in Urban Governance*

The formulation and effective functioning of a ward committee can be considered as an effective step to ensure participation of common people in the urban governance process. West Bengal is one of the few states where ward committees have been functioning for nearly a decade. The government of West Bengal has always been committed to decentralized and participatory urban local governance, through making ward committees more broad-based and participatory. The West Bengal Municipal Act of 1993 (Section 417) first made the provision for the formulation of ward committees. On the basis of that recommendation, the Department of Municipal Affairs of the Government of West Bengal has developed the West Bengal Municipal (Ward Committee) Rules (2001) to make the constitution of ward committees mandatory for all ULBs. This is done to ensure participation of people in the development activities of the cities, as well as in the urban governance.

Enhancing participation of poor people in cities does not simply mean empowering the poor to assist the powerful in advocating for physical rights. It also includes the responsibility to devolve power over decision-making to those most affected by the problems (Sridharan 2008). The participation of the citizens can be ensured at two levels of the city governance—planning and implementation (Ghosh et al.

1995). At the planning stage, citizens' participation ensures area-specific needs of different parts of the city, that is, the demand-driven approach to city planning and participation in implementation ensures the quality of services.

The ward committees have a range of supervisory and monitoring powers within the ward, and the committee has dual accountability—they are accountable to the people, as well as to the urban government (Datta 2003). Ward committees are chaired by the councillor of the respective ward and are constituted by two types of members—nominated by the councillor and nominated by the Mayor/Chairman-in-Council. These members are nominated from different sections of the society inhabiting a ward under the discretion of the councillor, as a result of which the party members or dedicated supporters of the ruling party in the respective wards get nominations. Therefore, the ward committee does not represent the voice of all the people in a ward. Although ward committees are meant to be non-political entities, they function as the political organization at the ward level. The councillor even has the power to remove a member from the committee, which indicates the overwhelming power of the elected representative. Therefore, any voice raised against the councillor's decision remains either unheard or ignored.

It is required to convene annual general meetings of the citizens every year and a monthly meeting at the committee level. In this annual general meeting, the councillor has to face the citizens of his/her ward by means of which the ward committee can make the councillors more accountable to the local people. The field-level observation over more than a decade has noted that during the earlier regime of the Communist Party of India (Marxist)-led government in West Bengal, the ward committees in the cities were very regular at organizing the annual general meeting, but the committee level meeting was not done regularly (once a month). The frequency of such meetings varied from once in 2 months to twice in a year. This difference at ward levels depended mostly on the levels of initiatives taken by the Chairman (ward councillor) and the secretary of the ward committees. The resolution adopted in the general annual meeting used to be sent to the municipality. However, our enquiry into the status of submission of the copies of resolutions taken in the ward committee's general annual meeting found that in most cities they were very few. This raises the question of effectiveness of the ward committees. The recent observations from the cities indicate that the ward committee meetings have become very irregular, especially during the new regime of the Trinamool Congress (TMC)-led state government since 2011.

Our household-level field survey indicates that poor people, mostly from slum areas, participate regularly in the ward committee annual meetings. Observations from our group discussions with the middle- and high-income group people in the cities regarding participation in city governance indicate a very low level of participation in the ward committee meetings and their functioning. However, from the in-depth survey, it has been found that poor people, in most cases, only attend the meeting on the direction of the local councillor and the other party members. They often confuse the ward committee meeting as a party meeting. In response to our queries on the compulsion of attending these meetings, they replied that if they do not attend the meetings, they would not get the benefits given to the poor people that come from the local government. This simple answer explains how political parties control the local governance in the cities of West Bengal. Although there are a number of problems in basic services, especially in water supply and sanitation, none of these issues are raised by them in the meetings. Moreover, even if they get to raise their local problems regarding basic services, no action is taken to resolve them. Therefore, they have stopped raising their voice and now remain silent in the meeting. A deeper analysis of the nature of participation reveals that what is happening is simply inclusion of more people, rather than enhancing participation in the decision-making process.

2.7 Financial Reforms and the Status of Municipal Finance

Strengthening municipal finances and promoting fiscal autonomy of the ULBs were among the major objectives of the 74th Constitutional Amendment Act (CAA). This act also mandated the establishment of State Finance Commissions to advise on the devolution of state resources to the ULBs and the management of financial resources in the ULBs (Datta 1995). The Government of West Bengal formed two Municipal Finance Commissions (the first in 1979 and the second in 1990) for recommendations on municipal management and finance prior to the 74th CAA. After the 74th CAA, the GoWB formed four State Finance Commissions (the first in 1995, the second in 2000, the third in 2008 and the fourth in 2016) and the Municipal Administrative Reforms Committee in 2001. All these committees recommended the strengthening of decentralization, transparency and the improvement of municipal finances, and they put a major thrust on the enhancement of individual income of the ULBs, to lessen their very high dependence on grants from the state government. The State Finance Commission reports give yearly details of municipal funding of each and every ULB. It is mandatory for all the ULBs in West Bengal to submit the Annual Administrative Report.

Accounting reforms have been implemented in all the ULBs in West Bengal in different phases since 2006 to correctly reflect the financial position of the ULBs, and to make their projects bankable and viable. This financial reform has been realized under the externally funded KUSP Programme. The KUSP put a lot of emphasis on government reforms, as a result of which the financial credibility of the ULBs has increased to a certain extent. In order to bring clarity in the municipal financial management system, the old single-entry system of accounting, that was being followed by the urban local bodies, has been replaced by the double-entry accrual-based system.

In spite of financial reforms such as increase in property tax, introduction of water tax and so on, the ULBs of West Bengal are still highly dependent on government grants. One of the reasons for this higher level of dependence lies in the municipal fiscal structure in West Bengal. Here the pattern is of property tax as the main municipal tax, along with major dependence on state transfers. Therefore, the ULBs are financially highly dependent on the state government with their meagre source of own revenue generation. The common reasons which are responsible for the imbalances in revenue and expenditure of the ULBs in India are defective pricing policy, limited revenue sources, lack of revenue efforts, obsolete norms governing price fixation, increasing marginal costs, etc.

From the field research, it has been noticed that a number of initiatives were taken to increase the income from own sources. These include enhancement of property tax, property tax collection of current demand, arrears and outsourcing of collection, advertisement tax through hoardings, establishment of parking zones and introduction of parking fees, implementation of water charges, enhancement of administrative fees, user charges for different services and leveraging idle assets. Although the tax collection rate has considerably increased in the recent past, the property tax base has not yet become a dependable source of revenue. This is sometimes due to the low rate of collection of both the current and the arrears of the tax. To recover the gap in property tax collection, ULBs are reforming property tax with the help of GIS in the cities of the state in general and in the four studied cities in particular. GIS technology is being used by the ULBs to identify the un-assessed properties' tax assessment and to increase collection efficiency. Introduction of the Property Title Certification System is also being done in many cities.

In spite of several financial reforms, municipal governments are not yet able to recover the cost they incur in providing different services. Kundu (2003) opines that it would also be erroneous to expect small and medium towns to finance capital expenditure through internal resources or borrowing from the capital market. That is why they are still dependent on state government grants (e.g. see Table 2.3). Although the increase in property tax and the introduction of water tax has increased the revenue income from own sources, yet the overall income has never reached even 50% of the total income in case of many ULBs. In recent years, there has been a considerable decrease in the revenue income from own sources in the cities under study. The figures in Table 2.3 between the year 2010–2011 and 2015–2016 clearly indicate this trend. This has happened due to the withdrawal of the water tax in the cities of West Bengal by the AITC-led new state government which has been in power since 2011, and it indicates how the financial condition of the cities, even their own income, is controlled by the state government.

The analysis of the Municipal Budget gives us a certain idea about the nature of urban services in the ULBs. Capital expenditure may be viewed as providing pure

Table 2.3 The proportions of own fund and government grants in the total revenue income of Burdwan municipality

Sources of revenue income (Rs.)	1988–1989	2003–2004	2007–2008	2010–2011	2015–2016
Own fund	7,977,997 (25.06%)	49,254,882 (37.95%)	102,478,803 (41.89%)	124,485,310 (33.43%)	158,470,503 (31.47%)
Government grant	23,854,451 (74.94%)	80,524,355 (62.05%)	142,183,460 (58.11%)	247,808,063 (66.57%)	345,057,148 (68.53%)
Total income	31,832,448	129,779,237	244,662,263	372,293,373	503,527,651

Source: Copies of the Municipal Budget

or impure public goods, whereas revenue expenditures most directly provide private goods such as office equipment, wages and salaries, and therefore, higher capital rather than revenue expenditures imply better service delivery (Sridhar 2007). The detailed analysis of the budgets of the four cities indicate that in each category of services, revenue expenditure is much higher when compared to capital expenditure, which means that these cities spend more on establishment and salaries than on development work.

2.8 Other Partners

2.8.1 Public–Private Partnership (PPP)

The state governments in India are no longer in a position to continue the subsidization of civic amenities, and in order to overcome the problem of government failures in providing basic urban services, they have been looking for unconventional methods of financing urban basic services (Bagchi 1999). Public–private partnership is one of those emerging unconventional methods of financing urban infrastructure and services. PPPs are cooperative arrangements between private and public sector organizations for providing infrastructure services and products to the public, where the parties agree to share duties, responsibilities, costs, profit, and risks. Association with the private sector is necessary to get technical know-how, financial resources and management skills. Stakeholder's association is significant to ascertain the user's preferences and demands, in order to reduce operation and maintenance costs of the scheme, to be able to ensure financial sustainability, and because of the fact that private providers use part of their profits to serve the poor (Bagal 2008).

To meet the increasing gap between the growing demand for infrastructure and services and the increasingly scarce public sector outlays, the government of West Bengal emphasized the public–private partnership initiatives in urban areas. The state government had adopted a policy on infrastructural development through PPP in 2003 and has just finalized 'the state government policy on infrastructure development through PPP in the urban sector' to be placed before the cabinet for approval.

Nurturing and implementing PPP projects are more challenging than conventional projects. PPP is often a complex arrangement and requires more skills. The ULBs in most cases are not capable enough to handle these projects. Therefore, the development authorities are taking up PPP projects instead of the ULBs. The majority of the PPP projects in the cities are housing projects. Health and solid waste management in many cities have also come under the PPP projects in recent years.

Sridhar (2007) opined that strong public sector governance can regulate public–private partnership projects to reduce the cost and to improve the quality of services. Otherwise, the private partner would take the profit and public sector would bear the risk. Some senior-level officers of the Urban Development Department of West Bengal have stated that in most cases, the private companies benefit more from the

PPP projects as the government does not have appropriate rules and regulations regarding the PPP projects. According to a news report (SNS 2009), the private partner plans, executes and completes the projects, whereas the state government officials are consulted only to complete the formalities. However, on the basis of the field observations, it is too early to comment on the effectiveness of the PPP projects in the cities.

2.8.2 Non-governmental Organizations and Community-Based Organizations

Successful urban management needs involvement of citizen's groups in the improvement of urban services. Among different citizen's groups, non-governmental organizations (NGOs) and community-based organizations (CBOs)[4] can play a significant role in the management of different services in the ULBs by providing support to the people, the community and the local government. The NGOs help mobilize local resources, communicate information about needs to local political leaders, lobby for programmes and policies to expand urban services, provide financial assistance, promote new approaches to service delivery and interact with the municipal officials, programme managers and field staff (Ghosh and Ahmad 1996). However, in comparison with the other states, NGOs got a very late entry in West Bengal as the state government vehemently opposed the NGOs to play their role as citizen's group in the 1880s and the early 1990s. Therefore, the role of NGOs in urban management and in service provision is extremely poor in the cities of West Bengal. From the field research in four cities in West Bengal, it has been noted that there are a few NGOs, especially in Burdwan and Durgapur, as individual service providers in the health and education sectors. Although they have an insignificant share in the total quantum of services in the cities, their services to the poor are nonetheless significant.

Though in the West Bengal Municipal Act much emphasis has been given to the participation of citizen's groups in the city governance, NGOs are still not involved in the processes in the cities studied. The NGOs in the cities remained as individual service providers and have no association with the local government. Although the health officers of the Burdwan and the Durgapur ULBs claimed that NGOs are invited under the health programme initiatives, it has been observed that they are only invited for preventive health care programmes like Pulse Polio, in order to get the assistance of more manpower in the effective running of such programmes.

[4] CBOs are non-profit organizations developed at the neighbourhood level to protect the interest of the local people especially in basic service provisions by the urban local bodies. In India, there are two kinds of CBOs: (1) developed independently by the neighbourhood committee such as residents associations and (2) developed by the local governments for better management of neighbourhoods.

Thus, it can be said that the NGOs could not develop in West Bengal as effective citizen's groups in urban governance.

For the urban poverty alleviation programmes, such as SJSRY, three levels of community-based organizations (CBOs) were created in the cities. These are Neighbourhood Group (NHGs) at the lowest level, Neighbour Committees (NHCs) at the second level and Community Development Society (CDS) at the city level. These different levels of CBOs are officially empowered with a number of responsibilities such as dissemination of information and medium of communication, encouraging people's participation, supporting planning, implementation and monitoring of activities at the local level, motivating Below Poverty Level people to join thrift and credit societies, facilitating recovery of loans and creating small community assets (Datta 2006).

However, empirical research at the city level observed that these CBOs are formed by the city government officials following the nomination of the local councillor. The councillor usually nominates people belonging to their party supporters. Therefore, these CBOs are not independent bodies who are able to raise their voices on different issues, as they are controlled by the party structure. All four cities under study do not have representative CBOs formed outside the local government institutions, and therefore, the role of CBOs is utterly negligible in monitoring the services and the overall governance process of the cities. From the empirical research at the city level, it is found that high political interference is one of the major obstacles in the participation of non-political institutions, such as NGOs and CBOs, in the activities of the city government.

2.9 Centralized Politics vs. Decentralized Governance

In spite of being the pioneer in decentralized urban governance and despite having early reforms and regulations, the ULBs in the state of West Bengal still suffer from centralized control of the political parties. This situation creates enormous problems in the way of participatory development in the cities. The election of ULBs is regular in West Bengal and takes place at an interval of 5 years, but the nature of elections is different from that of the other states. Here electoral process takes place directly under the political party, and the candidates contest under the party banner, whereas in many other states, such as in Bihar and Jharkhand, people are elected as independent candidates from the local area without direct involvement of political parties. Due to the involvement of political parties in the local government elections, local issues often do not get prioritized.

In West Bengal, since the election of local governments in the 1980s, the situation has been different. The dominant political parties of the state take a major drive to contest the local elections, and the central organization of the party decides which candidates would contest. The situation with the earlier government supported by the Left Front and the present government supported by TMC is no different in this context. According to the Municipal Act, the Mayor/Chairperson should be elected by the elected councillors. However, it was observed during field research that the decision is often taken by the central party authority so that the state-level organization of the party can control the local government as well as the city politics. From discussions with the councillors at the city level, it was found that they are ignorant about the way in which the Mayor/Chairperson is chosen in their own cities, and think that it is decided by the central party office located in Kolkata. In some states such as Jharkhand and Bihar, the Mayor/Chairperson is directly elected by people of the city, which is not the case in West Bengal.

To retain political control in the local government, the political party in power in the state tries its best to be elected, and this sometimes leads to violence during election of urban local bodies in West Bengal. The party in power in the state always tries to control the activities of the cities, and that is why there is always tremendous presence of political parties in the local governments. The government grants from the state to the city governments become skewed depending on the party colour of the local government. The cities run by opposition parties do not get equal share of government grants. It has also been seen in recent years that the elected members in the local governments change their party affiliation to ensure the grant to the ULBs and to keep the activities of the city running (Samanta 2020).

2.10 Critical Review of Governance Reforms in West Bengal

The decentralized model of urban governance in West Bengal has been efficiently running for decades starting from the period before the 74th Constitutional Amendment Act. People's participation has been mandatory since a long time in the state, through a number of committees such as the ward committee. However, empirical research shows that the formation and action of these committees are highly controlled by the political parties. The poor people are usually highly dependent on the city government, especially for water supply and other basic services. Access to those services is often controlled by the parties, especially for informal settlements. Therefore, the compulsion of the poor makes them participate more in the process of governance.

On the contrary, the middle class is not interested in the running of the local government as they do not depend on the municipal services in a big way. In small cities, people belonging to the middle class can have their own water sources within the premises of their own houses through deep tube wells, which are usually banned in big cities. They only depend on the municipal garbage collection system. Political interference in local governance is also a major reason why the non-political middle class do not participate in the governance process either through the ward committees or through resident associations. NGOs at the city level are also not found to be actively engaged in the governance process. The absence of bodies questioning the city governments' activities leads to excessive control by political parties. Thus, effectiveness of decentralized governance in city management is not found in a big way at the city level.

The most significant problem of governing small cities comes from the municipal funding. The Central Government offers grants through different urban development programmes, but these are usually routed to big cities rather than the small cities. Both the Jawaharlal Nehru National Urban Renewal Mission (JNNURM) from 2005 to 2014 and the new scheme, namely, Atal Mission for Rejuvenation and Urban Transformation (AMRUT) starting in 2015 facilitate funding for bigger cities. The funding under these Central Government schemes for small cities is negligible. To adjust with this situation and to get more funding from the Central Government schemes, the West Bengal government has started merging smaller cities with existing municipal corporations (MC). For example, Rajarhat-Gopalpur has been merged with Bidhannagar MC, and Raniganj, Jamuria and Kulti have been merged with Asansol MC.

The small cities are highly dependent on state government grants, as they either do not receive much from the Central Government or are not financially efficient enough due to poor economic bases. The dependence on state government funding makes them more politically dependent on the state government. The basic services are often affected by the financial limitations of the local governments and by the delayed arrival of government grants. The funds from the state government grants are also controlled by fixed projects, and therefore the funds are often tied. Untied funds are very rare, which limit the independence of the city governments to spend according to their specific needs. The generation of individual revenue is the only way to overcome this constraint, and small cities still have to go a long way to reach that stage, as they can only depend on property and commercial taxes for their major sources of income.

References

Bagal S (2008) Managing public private partnerships. Econ Pol Wkly 43(33):23–26

Bagchi S (1999) Myths of empowering urban local bodies. Econ Pol Wkly 34(37):23–26

Basu K (2006) Kolkata urban services for the poor: achievements and challenges. Urban Manag 12:20–31

Biswas S (2003) Introducing Kolkata urban services for the poor (KUSP). Urban Manag 11:63–65

Datta P (1990) Urbanisation and urban governance in West Bengal. Institute of Local Government and Urban Studies, Department of Municipal Affairs, Government of West Bengal

Datta A (1995) Municipal reforms in India: comparative models and processes. Econ Pol Wkly 30(38):2395–2398

Datta P (2003) Towards good governance and sustainable development. Dasgupta and Co. Pvt. Ltd, Kolkata

Datta P (2006) Decentralisation, participation and governance and development: context and concerns. Kalpaz Publications, Delhi

Ghosh A, Ahmad SS (1996) Plague in Surat: crisis in urban governance. Institute of Social Science and Concept Publishing Company, New Delhi

Ghosh A, Ahmad SS, Maitra S (1995) Basic services for urban poor: a study of Baroda, Bhilwara, Sambalpur and Siliguri. Institute of Social Science and Concept Publishing Company, New Delhi

GOWB (2003) Urban West Bengal: a policy statement. Department of Municipal Affairs and Urban Development Department, Government of West Bengal, Kolkata
GOWB (2017) Notification for newly merged departments and creation of new demands. Government of West Bengal, Finance Department, Kolkata. https://wbxpress.com/wp-content/uploads/2017/02/1505-FB.pdf. Accessed on 20 June 2020
Hamid A (2004) 74th amendment: an overview. CCS Research Internship Paper. New Delhi: Centre For Civil Society. https://ccs.in/internship_papers/2004/2.%2074th%20Amendment_Areeba.pdf. Accessed on 9 Oct 2016
Harris N (2003June 21) Globalisation and the management of Indian cities. *Econ Pol Wkly* 38:2535–2543
ILGUS (2002) Urban West Bengal. Institute of Local Government and Urban Studies, Calcutta
Khasnabis R (2008) The economy of West Bengal. Econ Pol Wkly 43(52):103–115
Kundu A (2003) Urbanisation and urban governance: search for a perspective beyond Neo-Liberalism. Econ Pol Wkly 38(29):3079–3087
Kundu A (2011) Method in madness: urban data from 2011 census. Econ Pol Wkly 46(40):13–16
Ministry of Housing and Urban Affairs (MoHUA) (n.d.) Background of 74th CAA. Ministry of Housing and Urban Affairs. http://mohua.gov.in/upload/uploadfiles/files/74th_CAA13.pdf Accessed on 5 April 2020
Mitra S, Mukherjee K (2008) Integrated and inclusive development of urban West Bengal: the KUSP perspective. Urban Manag 13:31–54
Pradhan KC (2013) Unacknowledged urbanization. *Econ Polit Wkly* 48(36):43
Rao CN (2016) Urban governance in India. Kalpaz Publication, New Delhi
Roy I (2006) Representation and development in urban peripheries: reflections on governance in Ahmedabad suburbs. Econ Pol Wkly 41(41):4363–4368
Samanta G (2010a) The globalizing state, public services and the new governance of urban local communities in India: a case study of West Bengal. Unpublished research project report funded by Ford Foundation, New Delhi
Samanta G (2010b) Institutional reforms and changes in governance process: a case study of West Bengal. In G. Shah et al., The globalizing state, public services and the new governance of urban local communities in India: colloquium. Vikalpa, The Journal of the Indian Institute of Management, Ahmedabad, 35(1): 96–103
Samanta G (2013) Urban governance reforms and basic services in West Bengal. In: Markandey K, Srinagesh B, Lonavath A (eds) Challenges of urbanization in 21st century. Planning and governance. Concept Publishing Company, New Delhi, pp 361–377
Samanta G (2014) Politics and classification and governance in census towns. Econ Pol Wkly 49(22):55–62
Samanta G (2015) Small and medium towns in eastern India: the dynamics of the economy, mobility and governance. Unpublished research project report funded by Indian Council of Social Science research, New Delhi
Samanta G (2017) New urban territories: transition, transformation and governance. In: Denis E, Zérah M-H (eds) Subaltern urbanization in India: an introduction to the dynamics of ordinary towns. Springer, New Delhi, pp 421–442
Samanta G (2020) Decentralized governance versus state dependence: financial challenges and participatory development in small cities of West Bengal. In: Thakur RR, Dutt AK, Thakur SK, Pomeroy GM (eds) Urban and regional planning and development: 20th century forms and 21st century transformations. Springer, New York, pp 321–336
SNS (2009) State finalises PPP policy. The Statesman, 19 April, Kolkata
Sridhar KS (2007) Reforming delivery of urban services in developing countries. Econ Pol Wkly 42(33):3404–3413
Sridharan N (2008) Globalisation of urban India. Econ Pol Wkly 43(10):26–31
World Bank (2006) Reforming public services in India: drawing lessons from success. Sage Publications, New Delhi

Chapter 3
Women's Empowerment Through Participation in Urban Local Governance

"Reservation for women in local government is not just a question of getting certain number 'in'; it also nurtures their capabilities to play their rightful roles in the development process and makes an important difference as the local government deals mostly with the quality of life for citizens." – Sharma 2015: 30

3.1 Women's Empowerment

Women's empowerment issue is inherently a complex one, as the issues related to women are varied and multidimensional (Samanta 2009: 2012). In its simplest form, women's empowerment urges for the redistribution of power which challenges patriarchy and male dominance. It acts both as a process and as the result of the process. It is transformation of the structures or institutions that reinforce and perpetuate gender discrimination. It is a price that enables women to gain access to and control of material as well as information resources (Avasthi and Srivastava 2001). According to Tapan (2000), empowerment of women means making their own choices with regard to their lives and makes them active players in the society. Gaonkar (2004) conceptualized the empowerment of women as a multidimensional and holistic approach. According to her, although it covers social, political and economic aspects of women's development, economic empowerment is of utmost significance in order to achieve lasting and sustainable development of society.

A number of studies consider three dimensions to be important for the empowerment of women. These are 'power to', 'power with' and 'power within'. The 'power to' dimension indicates the power of women to control their lives. The 'power with' indicates the collective power of women to negotiate their gender, caste, class and other interests. The 'power within' indicates strategic gender awareness. However, Acharya and Ghimire (2005) identified another three dimensions of women's empowerment: economic, social and political. According to them, each component

R. Banerjee, G. Samanta, *Negotiating Terrain in Local Governance*, Local and Urban Governance, https://doi.org/10.1007/978-3-030-60663-3_3

of these dimensions reinforces the other. The economic aspects include increasing women's access to and command over tangible and intangible resources, such as wealth, property, employment, knowledge and information. Social aspects include changing discriminatory ideology and culture, which determine the environment for women's existence. Finally, political process must increase women's presence and influence in the power structure.

However, Garba (1999: 131) adopted two genetic concepts of dimensions of empowerment with respect to women: the static and dynamic. The former emphasizes women's empowerment in terms of their capacities to participate in decision-making that directly or indirectly affect their lives and to influence those decisions. This refers to the notion of women having an effective voice. Women are also assumed to be disempowered when they cannot influence decisions that alter their lives. The static concept of empowerment leads to exogenous empowerment strategies. Exogenous strategies are those built on the premise that disempowered groups can be empowered by external individuals or groups in a way that an effective voice could be given to women. The exogenous empowerment strategy implies a top-down approach.

In contrast, the dynamic concept regards empowerment as a process of developing the capacity of individuals to participate effectively in making and implementing decisions that directly or indirectly affect them. Viewed as a process, empowerment is something an individual or a group of individuals acquires over time. It is not something you can give to people, although the conditions could be created to increase their chances of acquiring it by themselves. The dynamic concept of empowerment leads to endogenous strategies. Endogenous strategies are those whose underlying premise is that external groups can only facilitate empowerment by creating enabling conditions for disempowered groups to empower themselves. The endogenous empowerment strategy is associated with a bottom-up approach to women's empowerment. However, exogenous forces can enable or accelerate the endogenous process of empowerment.

Karl (1995) offers a useful framework for identifying the exogenous enabling conditions for endogenous empowerment, describing four stages in the empowerment process with respect to women: (a) awareness; (b) capacity-building and skills development; (c) participation and greater control in decision-making; and (d) action for change. Garba (1999) extends Karl's (1995) four-stage process into a six-stage empowerment process: (a) awareness; (b) skills and capacity assessment; (c) capacity-building and skills development; (d) participation and greater control in decision-making; (e) action for change; and (f) evaluation.

Hashemi and Schuler (1993) hypothesized that the process of empowerment can be attained by weakening the systemic basis of women's subordination. According to Batliwala (1994), women's empowerment can be attained through the process of:

- Challenging the ideology of male domination an women's subordination
- Enabling women to gain equal access to and control over resources

- Transforming the institutions (family, education, religion, media, etc.) and structures (legal, political and social) through which the ideology and practice of subordination is reinforced and reproduced

On the types of changes favouring women's empowerment, Cheston and Kuhn (2002) stated that although the process of empowerment varies from culture to culture, several types of changes are considered to be relevant in a wide range of cultures. Some of these changes include increased participation in decision-making, more equitable status of women in the family and community, increased political power and rights and increased self-esteem. According to IFAD (2000), four broad categories of processes of change can lead to women's empowerment. Those are (a) change in women's mobility and interaction, (b) change in women's labour pattern, (c) change in women's access to and control over resources and (d) change in women's control over decision-making.

Attempts to empower women—socially, legally and politically—cannot be divorced from the social and historical reality of their status and position in society. According to Tapan (2000), the issue of women's empowerment cannot also be isolated from the issue of poverty, violence against women and child abuse. Charmes and Wieringa (2003) highlighted the importance of two aspects of the concept of women's empowerment. According to them, the first aspect of empowerment should deal with it as a field of operation, its dimension, its interlinkages as well as its intersectionalities with other fields of power relations, such as those of race/ethnicity and class. In the second place, women's empowerment can be seen as a process in which the following elements will be considered: awareness/consciousness, choice/alternatives, resources, voice, agency, and participation.

Lastly, it can be said that the process of women's empowerment highly depends on the existence of alternatives. Women may be aware of the conditions of their oppression, but if they see no viable alternatives, they can only turn their anger into frustration and bitterness or into acceptance of suffering. Thus, Women's capability to make meaningful decisions over critical areas of their lives depends to a large extent on the existence of alternatives to arrangements about which they are unhappy (Charmes and Wieringa 2003).

3.2 Women's Development and Empowerment Programmes in India

The representation of women in developmental interventions in almost all countries is not free from gender bias. The interventions were highly in favour of males, and it was assumed that women would automatically benefit from improvements in the conditions of their families or of their male counterparts. However, in reality women remained recipients, whose major roles were at home, raising the children and minding the household, rather than participating in the process of development. As a result, they become powerless, which is rooted in four basic structures: those of production, reproduction, sexuality, and socialization of children (Waylen 1996).

In developing countries, the productive roles of women are not acknowledged, rather they remain invisible (Boserup 1970). Women are always perceived as mothers and/or housewives; their needs are determined to be maternal and child health and nutrition. The national or state machineries for women, ministries and government departments, in almost all third-world countries, reflect this approach to women's issues in their nomenclature; in India, for example, the relevant ministry is still known as the Ministry of Women and Children's Welfare. Similarly, we see the Malaysia Ministry of Women, Family and Community Development, and in Bangladesh, it is named Women and Children's Affairs, putting women and children's concerns in the same category. However, in recent years, there have been a few changes, and the Pakistan government now calls it the Ministry of Women's Development, Indonesia has the Ministry of Women's Empowerment, and Sri Lanka has a Ministry of Women's Empowerment and Social Welfare.

In the developmental efforts of the first two decades of planning in independent India, no special emphasis was given to the development of women. As a result of the male-biased development intervention, women's position in society changed in a negative way with the improvement of the economy as well as social status of their family. In the poor and low-income families, women have access to valued productive activity which is often invisible in the middle-class families. Since the 1970s, many studies, especially the 'Towards Equality' report in 1974, started to highlight the gender discrimination and consequent impoverishment of women, their low status despite growing work burdens, their oppressions and their invisibility in the productive economy, as well as in development programmes (Samanta 2016a).

To remove the gross inequalities between men and women in society, the Indian government started to take up separate policies and plans for women from the mid-1970s (Madhok 2003). This was partly due to the international pressure created by the United Nations and partly as a response to the 'Towards Equality' report of 1974 (Rajagopal and Mathur 2000). The policies and programmes are many in number, among them the followings deserve special mention:

(a) National Plan of Action for Women – 1976
(b) National Nutrition Policy – 1983
(c) National Health Policy – 1983
(d) National Policy on Education – 1986
(e) National Policy on the Empowerment of Women – 2000
(f) Institutional Support:

- An Exclusive Department of Women and Child Development – 1985
- National Commission for Women – 1992
- Rashtriya Mahila Kosh (RMK)
- Parliamentary Committee on Empowerment of Women – 2001

In the above mentioned policies and programmes, women were perceived by the state as mothers, and their needs were determined to be maternal and child health and nutrition. Another important form of representing women is to put them in the same category along with children. By setting up an exclusive department for women and child development, the state conforms to the gender roles of women as mother and caregiver (Samanta 2016b). However, more important issues for women are education and jobs. There is no doubt about the fact that lack of education and gainful employment for women and the consequent economic marginalization are the root causes of violence against women.

A range of literature has emerged portraying women as a homogeneous group of resource users burdened by the subsistence needs of their families, as victims of degradation and poor availability, with low access to resource ownership, and as the 'natural' conservators and nurturers of local environments. It is true that gender inequalities are a reality of life across India, evidenced by almost every demographic indicator and a host of social and economic statistics. It is also generally true that throughout India, women are burdened with the chores of collecting fuel, fodder and water, trying to help their families to survive. The marginalization of women in the post-colonial Indian state has gradually driven them into the most insecure and informal jobs (Lahiri-Dutt et al. 2006).

The analysis of the details of each and every programme meant for women in India till the last decade of the twentieth century clearly indicates that the real empowerment of women in the true sense of the term was not the priority of the state of India. As discussed in the earlier section, empowerment of women is closely associated with two basic factors which are again linked to structural changes. These two factors are (a) more control over access and distribution of resources and (b) reorientation of the existing power structure. To facilitate the one, India needed to guarantee the second. Here comes the need for women to enter into the political–economic power structure, through the democratic process of election.

The last decade of the last century became significant in the lives of Indian women as many things were happening in favour of their empowerment. The National Perspective Plan for women (NPP) was developed in 1988, which discussed the issue of political participation and under-representation of women at all level of governance. Following that, the 73rd and 74th Constitutional Amendment Acts were passed to ensure one-third reservation of women in local governance. The 73rd Constitutional Amendment Act was for rural local government called Gram Panchayats in India, and the 74th was for urban local bodies. There were four states (Kerala, Karnataka, Andhra Pradesh and Maharashtra) in India, which had reservation for women in the local government even before the 73rd and 74th Constitutional Amendment Act were passed (Oomen 2014). It was not very difficult to pass the reservation for women at the local government level across India, and all the states implemented the same without much delay. Even the parties in opposition, with strong male chauvinist decision-making structures, supported reservation for women at local government level.

However, when in 1996 the Indian government wanted one-third reservation for women in the lower house of the Parliament and in the State Assemblies, it was opposed by all parties. Since then, the bill for reservation of women in upper strata of governance has been placed many times, but each time it was severely objected to by all political parties dominated by men. Even now, it has still not been passed in the Indian Parliament. The arguments for opposing the bill were many; and most of those were baseless. Some parliamentarians said that the Indian Constitution guarantees equal rights for men and women, so why do we need reservation for them in the Parliament? Another argument was that women will be the proxy of men, and it will be the government of *Biwi-beti-bahu* (wife-daughter-daughter-in-law) brigade. Some parties even demanded quota within quota, which means that with the reservation of women, there should be further reservation for women from Scheduled Caste, Scheduled Tribe and Other Backward Castes. According to them, otherwise, only 'short-haired urban women' will come to the Parliament, and poor lower-class women will never get entry. There was also a difference of opinion on the nature of reservation for women. One group demanded that all parties should have one-third electoral candidates as women, whereas the other group said that reservation of women should be of one-third of electoral seats. Thus the debate is still on, and we feel that it will continue for a long time, since it continues to create a crisis for the male members of the Parliament who seem to feel that by passing the bill they will not get the chance next time to become electoral candidates in their own constituency.

In the 1990s, many other things also happened around the world to reinforce the demand for women's access to resources and power. During the 8th 5-year plan period (1992–1997), gender sensitization through education, women's employment and skill training for women were given priority. The Indian government also had to prepare a country report for women to place before the Beijing Conference conducted in 1995, where the Indian government admitted that women's empowerment and participation are a political process linked to changes in power relations, and that patriarchal control in all aspects of life severely limits the potential and scope of women's actual participation in public life. The government of India prepared a Plan of Action for women for 1998–2000, and set up the Shram Shakti Commission to evaluate the function of women in the informal sector. The 9th 5-year plan (1997–2002) tried to develop a link between social change and development and portrayed women as the agent of social change. In India, the National Policy for the Empowerment of Women was initiated in 2001. Women's entrepreneurial role was recognized in the microfinance movement, and the issue of women's empowerment received greater attention from the policy-makers, especially of Microfinance Institutions (MFIs) in the development discourse of the recent past (Samanta 2009). It has also been observed that women engaged in microfinance and self-help group activities are also entering into politics and local governance in West Bengal (Samanta 2018).

Despite a long history of such plans and programmes since independence, Indian women still remain marginalized. In the early days, plans borrowed from and imposed by foreign organizations left no scope for Indian women to keep pace with the increased efforts for women's emancipation adopted to reach the point of gender equity (Tapan 2000). According to the Gender Inequality Index 2017, India's rank in gender equality is 127 out of 160 countries of the world. The maternal mortality rate is still 174 per 1,00,000 women. Only 27.2% women are in the labour force,

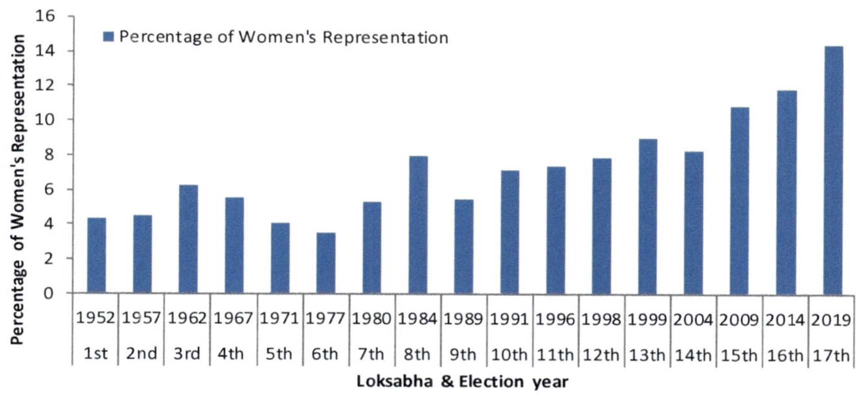

Source: Election Commission of India

Fig. 3.1 Percentage of Women's Representation in the Lok Sabha: 1952–2019

against 78.8% men. Following the election in 2019, only 14.5% women are in the Indian parliament, which represents an abysmally low participation of women in higher level of politics and governance (Appendix 3.1 and Fig. 3.1).

3.3 Women and Urban Local Governance in India

The Constitution of India (1952) had promised the Indian citizens to secure their social, economic and political justice and the equality of status and opportunity. However, Rai (2011) has argued that women still remain excluded from productive resources, property rights, and even in different spheres of life. Omvedt (1987) has explained the effect of this exclusion on women. According to her, exclusion has certainly increased the helplessness among women to combat atrocities and violence, whether coming from within the family, social groups, private *goondas* or the state and police itself. Omvedt (1987) has also narrated a brief history of women's struggle in decision-making through her studies in India.

The struggle of women to be part of decision-making bodies had started during the 1970s and 1980s, through a vigorous women's movement. It may be for environmental protection such as the Chipko Movement in Uttarakhand, or to protest against *bossism* and corruption in Maharashtra (Omvedt 1987: 1992). In both cases, women raised questions to the representatives of local government, especially at the village level. They asked why the representatives and members of local government were only men and not women.

Mazumder (1989), Omvedt (2005), Ghosh (2009) and Rai (2011) have highlighted the report of Committee on the Status of Women in India (1974) to understand the conditions of women during that time. In 1974, the report of the Committee on the Status of Women in India (CSWI) entitled 'Towards Equality' highlighted the worsening condition of women in India. It had also been interwoven with the purpose of the movement. In 1976, the published report of CSWI had suggested that women's representation at the grass-roots level political institutions was the need of the hour and should be increased through a policy of reservation of seats for women. After publication of the report, a national-level consensus has developed on the demand for reservation of seats for women.

In the history of the struggle for women's access to resources and governance, Maharashtra made a significant contribution as narrated by Omvedt (1987). In the late 1970s and early 1980s in Maharashtra, women had developed a women's panel with the Indian National Congress party for village-level elections. In 1984, a freedom fighter named Usha Nikam joined the women and submitted the women's panel for Gram Panchayat elections. After that, the family members of those women who were in the panel faced enormous oppression in terms of provision of basic services like water. Their husbands were threatened by the male political leaders. Due to such excessive pressure, the entire original panel of women, except for the name of Usha Nikam, was withdrawn. Then Usha Nikam developed another panel of women who could withstand the pressure, drawing from the Dalits, Muslims and lower-caste communities (Omvedt 1987: 1992). This movement gained publicity through some articles and journals. As a result, the idea of women's political power began to spread.

In November 1986, the Shetkari Sanghatana was developed as an independent farmer's organization, and thousands of peasant women and men attended the seminar organized by them. In the conference, they raised their voice opposing violence against women and advised the women to unite to secure their power (Omvedt 2005: 4748). However, this was totally ignored. After that, Shetkari Sanghatana developed training camps to prepare women for the elections of the Zila Parishad in 1987. The manifesto was prepared to provide all the basic services, as well as security to the women and their families, who would join the training. Other organizations also gave support to Shetkari Sanghatana. A number of individual women activists became quite influential, from a wide variety of backgrounds, and also promised to participate. This election was interwoven with the uncertain political situation of the country (Omvedt 1987: 1993), as V.P. Singh (former Prime Minister of India) developed a *janmorcha* in the country.

This incident provided mass awareness among the women of the country. During this period, reservation became the main demand in order to bring equality of status to women. However, at that time, caste issues became more prominent than the gender issue in the context of reservation (Sharma 1998; Ghosh 2009). As a result, the women's movement could not present a united front, and besides, the impetus for pushing for reservation came from men in the parliament. In 1988, 'The National Perspective Plan for Women' recommended 30% reservation for women in Panchayats and in Municipal bodies. By the end of the 1990s, several political parties had also made this demand. This led to the adoption of the 73rd and 74th amendments to the Constitution in 1993 (Omvedt 2005; Ghosh 2009; Rai 2011).

These amendments were passed by the Indian Parliament to usher in a new era of enhanced autonomy for marginalized groups of women, who have remained largely voiceless, passive and almost invisible citizens. Described as a 'ground-breaking' legislation by some, the result has been that millions of women have gained some experience in local politics. Sathe et al. (2013) have identified the main features of the 74th Constitutional Amendment as (a) direct election of women to all seats of the municipality; (b) reservations for Scheduled Caste, Scheduled Tribe and women; and (c) 33% reservation for women for the seats of local representatives, and if the states so desired, they could make provisions for reservation to the membership for Chairpersons to citizens of backward classes.

For urban governance, the 74th Constitutional Amendment Act of 1992 is significant because it reformed the Municipal Councils by seeking to provide power to the people, and by granting constitutional recognition to a third tier of urban local government. The Act provides for several changes in the structure of urban governance, its composition, its power and functions, but the primary interest to us is that this amendment was meant to empower women in the political and administrative domains in urban areas, which was till then a masculine arena with only limited exceptions. The reservation of seats created a political space for women in the male-dominated political milieu in India. This enhanced political participation was supposed to lead to social justice, equality of status and political empowerment of women. Indeed, for many women political representatives in small cities, just being able to attend the meetings was a major step towards equality. As a result, family relations have changed in many cases, and some women have become role models for others (Lahiri-Dutt et al. 2006). However, as the small and big city disparities in India are wide, the two contexts reflect different status and position of women. Women's education, emancipation and reformist movements in some metropolitan cities in India have a long history. Consequently, often women have emerged as excellent and able leaders in the metropolitan cities.

Now, the question is how the women in small and medium cities reacted to their becoming part of the decision-making process through a change initiated from above. We are attributing significance to the 'local and particular' in every way, the place itself being neither rural nor metropolitan, with a distinctive history and local culture, and with specific environmental issues that are rooted in local geography and economy (Lahiri-Dutt et al. 2006). We reveal the case of medium-sized cities in West Bengal, a Left Front-dominated state in India, where not all women councillors are completely new to politics and have reasonable levels of education. The perception of women councillors in urban governance embodying freedom and decision-making actually shows how the generalization of all women into one group can be a dangerous construction.

As we will show, women are a minority in urban governance and in decision-making bodies within the municipality. Thus, they lack critical mass in the political arena, which is necessary to develop any agenda as 'their own'. Again, many of the women participating in urban governance come from middle-class backgrounds of a politics that itself has been dominated by the middle classes. In this book, we shall explore how despite the changes in the Constitution the impacts are not straightforward; they have been complex, whether in terms of the impact of women on local urban government and political processes or vice versa. Above all, just by virtue of being 'a woman' does not mean their perceptions of power and decision-making would be different from men; middle-class women, in most cases, do not reveal particular awareness of gender concerns. This leads us to the problematic of women's awareness of the political and governance process and the hidden inequality within the power structure, even at the local level.

3.4 Women in West Bengal

West Bengal is one of the most progressive states of India especially in terms of the condition of women in comparison with many other states. This has historical roots that go back to the first phase of the feminist movement in India (1825–1915) which was started by Bengali educated liberals, including social reformers such as Raja Ram Mohan Roy and Ishwar Chandra Vidyasagar. Whereas Ram Mohan's struggle was for abolition of social discrimination such as Sati (burning of wife alive along with the husband's dead body), Ishwar Chandra Vidyasagar fought for access to education for women and widow re-marriage. They were the pioneers who sought to change women's lives in India, and specifically, in the then Bengal region. In the late nineteenth and early twentieth century, another liberal Hindu group came up in Bengal called the Brahmo Samaj. The contribution of Brahmo Samaj in changing the lives of women in Bengal in general and in Kolkata in particular was tremendous especially in terms of access to education and mobility rights of women to public spaces. The emergence of educational institutions for girls and women led to the birth of several women's organizations in West Bengal in later days. In 1882, Swarnakumari Devi established Sakhi Samaj (women's group) in Kolkata for the emancipation of women, and in 1901, Sarala Devi Chowdhury started Bharat Stree Mahamandal which was the first all-India women's organization. By the middle of the twentieth century, around 1946–1950, the Tebhaga movement took place in Bengal and North Bihar under the CPI and Kisan Sabha (farmer's association) for one-third share of the crop for share croppers. In that movement, a large number of women took part (Samanta 2016a, b).

The long history of the above-mentioned organizations and movements had a long-term impact on women's emancipation in West Bengal. Even now, if we look into the comparative data on a few gender parameters (Table 3.1), we can see that West Bengal performs much better than the all-India average on many counts in terms of women's performance. The FMR is higher than the all-Indian average. The level of literacy is higher by 5% points, bringing down the gender disparity in literacy. The workforce participation rate (WPR) is lower than the Indian average, but in India, it is mostly the agricultural labour force which increases the WPR for women. In West Bengal, women's agricultural labour force participation is 10% lower than that of all-India (Table 3.1). In contrast, the share of household industry workers is higher by 12%, and the percentage of female marginal workers is lower by 6%, both of which are better indicators for women.

Table 3.1 The comparative gender parameters in West Bengal and in India

Gender parameters	West Bengal	India
Female male ratio (per thousand)	950	940
Female literacy rate (per cent)	70.16	65.46
Gender disparity in literacy (per cent)	11.51	16.68
Female workforce participation rate (per cent)	23.13	31.11
Female agricultural labourer (per cent)	41.89	52.74
Female household industry workers (per cent)	74.03	62.77
Female marginal workers (per cent)	44.45	50.78

Source: Census of India 2011

3.5 Women in Urban Local Governance in West Bengal

After the 74th Constitutional Amendment Act (CAA) was passed in 1992, the West Bengal Municipal Act was developed in 1993, keeping the norms of governing cities in alignment with the 74th CAA. Previously the municipal affairs of this state were administered by the Bengal Municipal Act, 1932. Later, by the end of the 1980s and especially after the 74th CAA, it was felt necessary to replace the said Act with an updated legislation. Accordingly, the West Bengal Municipal Act, 1993 was enacted on 13 July 1994, and the new Act replaced the Bengal Municipal Act, 1932. Afterwards it has been amended 13 times, up to 2012.

Although the level of participation of women in Parliament and State Legislative Assembly in West Bengal is not much different that of other states (Ghatak 2010), the participation of women in local governance, especially in rural areas through the Gram Panchayats, was good in comparison with other states. Therefore, it can be said that West Bengal has a good record of participation of women in local governance, and that credit goes to the 74th CAA. However, if we try to look into the results of all the Parliamentary Elections in India with a gender-disaggregated database and compare the all-India case with West Bengal (Fig. 3.2), we can clearly see that the proportion of women representatives from West Bengal was always higher than the all-Indian average. This observation is probably in line with the fact that in the other gender parameters as described in Table 3.1, West Bengal is much ahead of Indian average figures.

Following the rules of the 74th Constitutional Amendment Act, the number of women councillors in any city or state should be one-third of the total number of councillors (elected representative from each ward of the city). However, while doing the field work at the city-level, we noticed that the actual numbers of councillors are more in number than expected by reservation (Samanta 2016c). To understand the present situation of women councillors in the cities of West Bengal, we have taken into account two major criteria: (a) observed and expected distribution of women councillors with respect to their reserved seats, and (b) political party-wise representation of women councillors in different districts of West Bengal.

3.5.1 The Observed and Expected Representation of Women Councillors in West Bengal

There are different types of arguments put forward by feminist scholars in reference to women's reservation in governance. Whatever be the argument behind it, it is well-known and accepted that to empower women and to fight the inequality from which women suffer on all counts, reservation for women in governance is needed. Although it has not been done yet at the higher levels of governance in India, the local level governance has ensured 33% reservation for women. However, in case of West Bengal, we have seen that even at higher levels such as in Parliament, the proportion of women representatives from West Bengal is much higher in comparison with other states (Fig. 3.2). At the local level, it has also been seen that the actual number of elected women representatives (EWR) is higher than the number reserved for them (Fig. 3.3). Taking into account all cities of all the districts together, the number of EWRs or councillors according to the reservation should be 878, but the observed EWRs are 1069.

In understanding the pattern of quantitative representation in greater detail, we looked into the district-wise pattern of difference between observed and reserved seats, as the city-level data including 129 ULBs is problematic to understand. In calculating this pattern, we encountered a problem, as the election of urban local

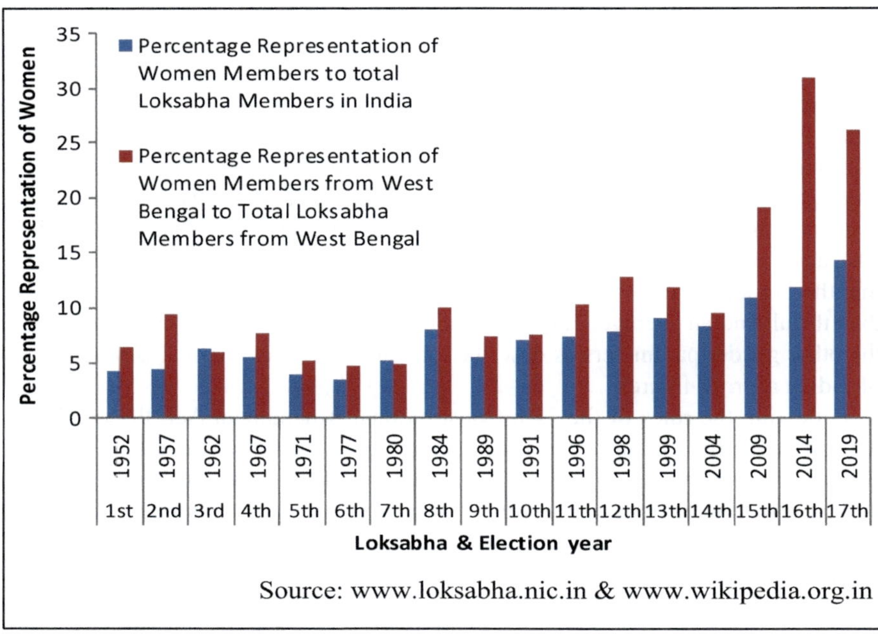

Fig. 3.2 The comparative representation of women in parliament from India and West Bengal: 1952–2019

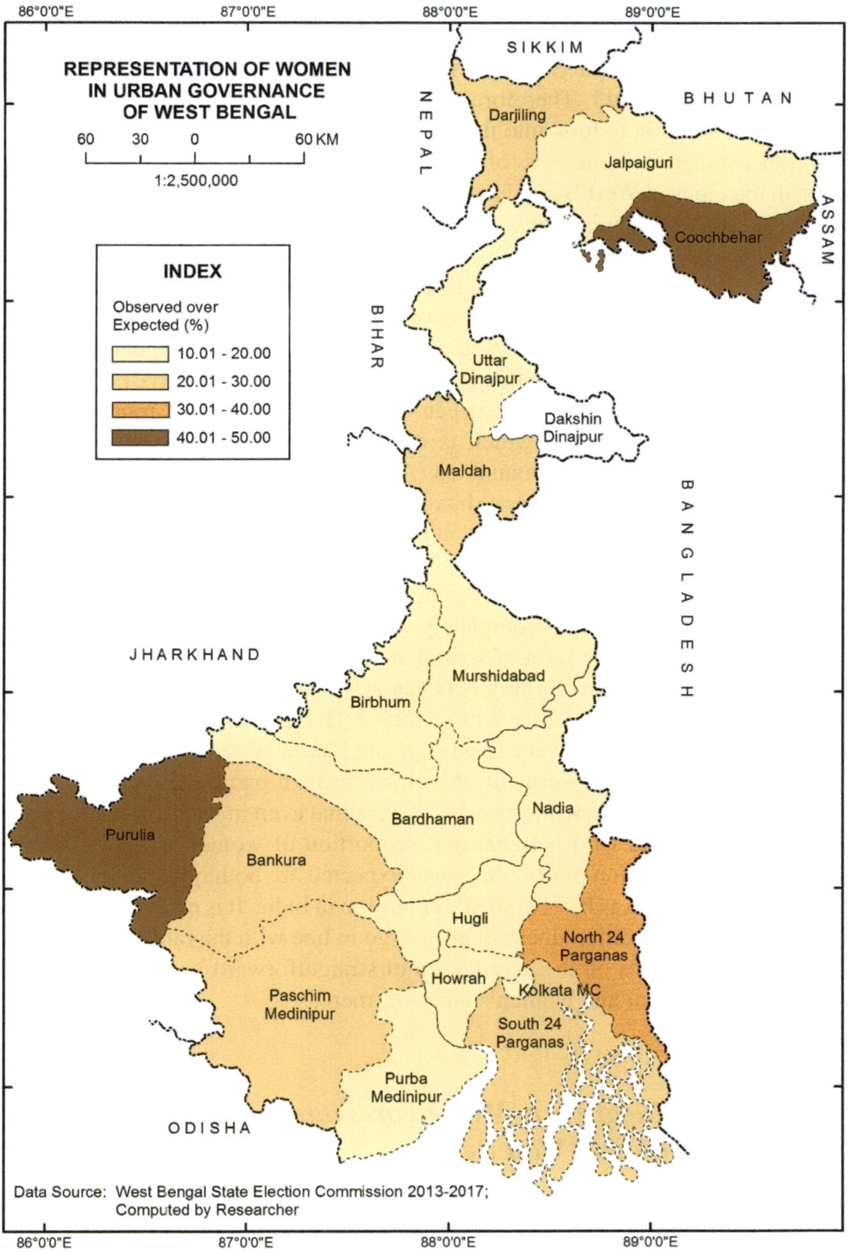

Fig. 3.3 The district-wise pattern of the observed over expected elected women representatives in cities of West Bengal

bodies in West Bengal does not take place at the same time, as it happens in case of rural local bodies called Gram Panchayats. To cover the data for all the ULBs from the website of West Bengal State Election Commission, we had to take a range of 4 years, from 2013 to 2017. Therefore, we had to comply with this problem as there was no other solution to look into this kind of data.

On an average, around 40% of elected representatives in urban local governments in the cities of West Bengal are women, whereas the quota for women is 33%. This shows that women's representation in urban local governance is more than expected by reservation. The district-level pattern (Fig. 3.3) is also significantly important. Out of 18 districts in West Bengal, 9 districts have EWR more than 40%, whereas the other half, i.e. 9 districts, have less than 40% representation of EWP. In each and every district, the observed value is more than the mandatory reserved percentage, except in Dakshin Dinajpur located in the northern part of West Bengal. A very high level of difference between expected and observed representation is found in North 24 Parganas, which is nearly a continuous built-up area and an extension of Kolkata Metropolitan Area (Appendix 3.3). The actual number varies with the number of ULBs in each district as well. In the districts that have more number of cities, there is a greater gap in between the observed and expected numbers.

However, if we look into the percentage of observed over expected (Fig. 3.3), it clearly shows that this data is completely in sync with the percentage of women in ULBs in that district. The level of spatial variation in the pattern of observed over expected is also very high, varying between zero in Dakshin Dinajpur and 43.75% in Pulia. Out of 19 districts, 7 have more than 25% of excess women representative over reservation, which is a very good sign and places West Bengal as a good state in terms of political empowerment of women through participation in local urban governance. The most interesting observation is that even in backward districts, like Bankura, Purulia and Cooch Behar, the proportion of women's representation in ULBs and the proportion of observed over expected are both quite high. This small observation speaks a lot about women's condition in India. It is evident that improvement in women's condition does not always go in line with the rate of development, which raises questions on the assumption of straightforward links between overall levels of development and women's empowerment.

3.5.2 The Political Party-Wise Representation of Women Councillors in West Bengal

The election of local government involves different policies in different states of India. In some states, the electoral candidates fight for election at the local level without any party banner, which means they get selected as local representative without having a political party colour, whereas in other states, the local level election is held directly under party banners. In the state of West Bengal, since the very beginning, the local-level election has taken place under the political party banner. Here the electoral candidates are selected by the political parties and are often supported by the higher-level party hierarchy. Therefore, an analysis of the proportion of women representatives under different political parties and their pattern over different districts helps us to understand the willingness of different political parties to induct women into the system of local governance.

This section looks into the proportional representation of women councillors under four different political parties under which the councillors contested the elections. The major parties contesting for election in West Bengal are the All India Trinamool Congress (AITC), Bharatiya Janata Party (BJP), Indian National Congress (INC) and Communist Party of India (Marxist). Among all these parties, till 2009–2010, all the cities were governed by the Communist Party of India (Marxist), and since then with the change of power in the state, the majority of the cities are under the AITC. If we look into the proportion of women's representation under different political parties, we can see that party-wise variation is also high besides district-wise variations (Appendix 3.4; Fig. 3.4).

Among the four political parties contesting for elections in different cities for the ULBs, the average West Bengal pattern varies but not very much. The AITC is the dominant party in power in more than 90% cities of West Bengal. They have around 44% women representatives in different cities. The second in line is the Communist Party of India (Marxist) with 37% women representatives. This political party was in power in different cities for three decades ending in 2010. They had a very active women's wing of the party at the local level. In spite of that, they have 37%, which is slightly higher than the expected 33%. The INC and BJP have 32 and 31% EWRs, respectively, at the local level. However, their overall representation is abnormally low in the cities of West Bengal.

If we compare different political parties' representation of women in different districts, it does not give us any clear pattern. However, from our empirical understanding of the situation in West Bengal, we can say that the district-level pattern depends on the political party's strength at the local level (Banerjee and Samanta 2017). In whichever district one political party is in a better power position, they will tend to have more EWRs. This might also hint at the fact that they want to get more women in politically safer areas; as in West Bengal, political violence is also very high in many districts.

3.6 Women's History in the Selected Cities

From the analysis of the present research, it has been found that the women councillors in Darjiling and Balurghat are in a more empowered position. In the cities of Raniganj and Hugli-Chinsurah, discrimination, opposition and proxy nominations are more commonly seen. Therefore, the empowerment level of the women councillors is very low there. While searching for the reasons for this differential level of empowerment of the women councillors in these cities, it was found that Darjiling

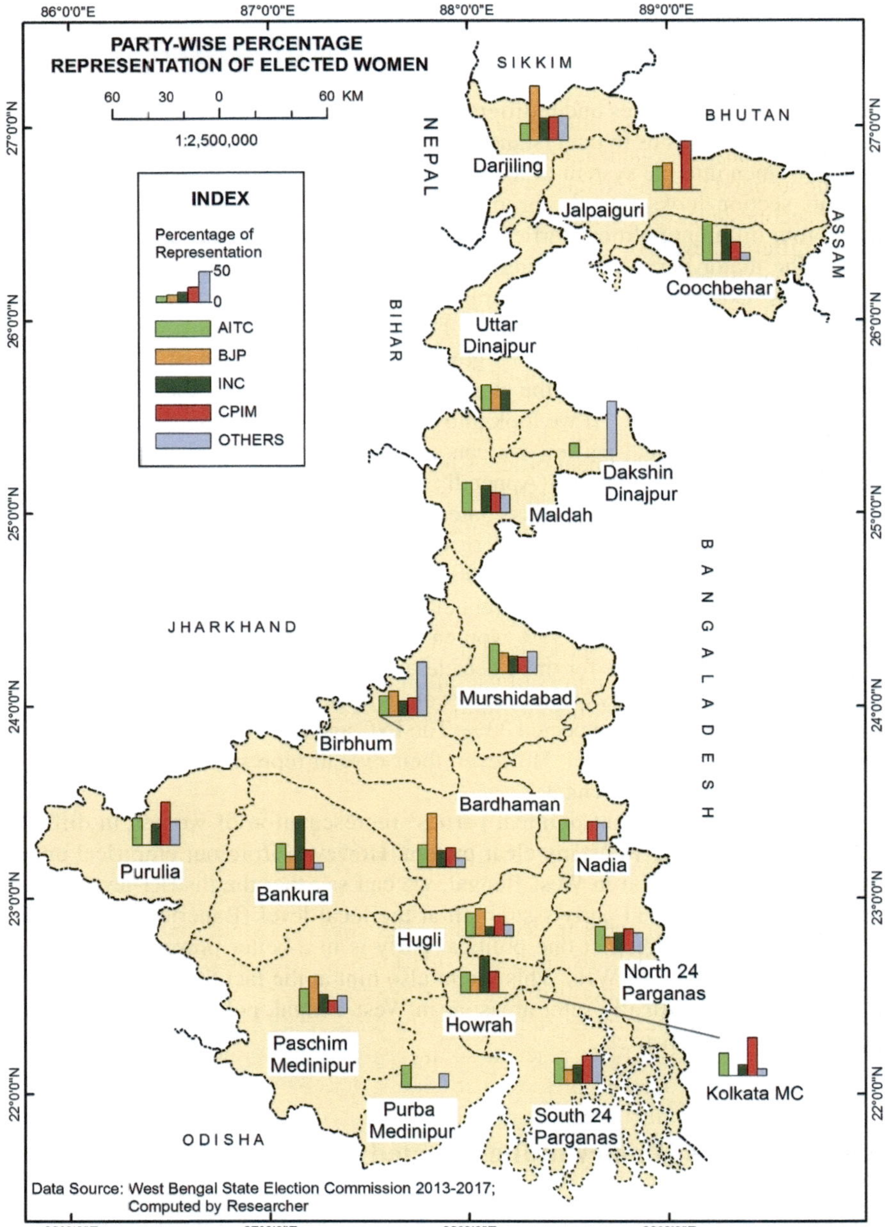

Fig. 3.4 The district-wise patterns of the party-wise representation of elected women representatives in cities of West Bengal

and Balurghat have a history of women's movement and women's struggle. That is why women are more empowered in these areas. This past history also positively influences the present emancipation status of the women councillors to participate in governance.

Darjiling has a very different context both in history and in geography among the selected cities of West Bengal. However, there is lack of sources to gain a comprehensive history of the city, especially the history of women's position in society. After extensive research from different sources including the local library, we were able to find one dissertation for the degree of Master of Philosophy on the history, political participation and empowerment of Nepali women in Darjiling, titled 'Women in Gorkhaland Movement: A Sociological Study' by Rai (2015). This section is largely developed from that thesis, supported by interviews with some elderly people, who were able to narrate some of the history for us. The position of women in Darjiling is rooted in its multicultural and multilingual regional character, with mixed communities of Nepali, Bhutia, Sherpa, Tibetans and so on, who have each brought vast changes in the social structure of Darjiling. Besides, the political movement for Gorkhaland also facilitated women's emancipation in this city.

In the traditionally rooted patriarchal society, Gorkha women enjoyed a respectable position in social and economic life besides their male counterparts. However, the situation was not the same earlier. They were also oppressed by hierarchical social structures, caste practices and gender inequalities. They were debarred from educational facilities. Rai (2015) has highlighted the history of the women's struggle in Darjiling in his study. The contribution of a few Gorkha women in this struggle is very important to mention. They were Putalimaya Devi Poddar, Mayawati and others who played an active role in dismantling the social structure by participating in the freedom movement equally with the men.

The Gorkhaland movement began with an attempt to preserve the cultural identity of the Gorkhas (Indian Nepalis) which they felt was threatened due to extensive immigration of people from Nepal to India during the colonial period. The movement began in 1907. However, it got popularized in 1986–1988 by Subhash Ghisingh of the GNLF party (Gorkha National Liberation Front). Then it stopped with the development of the Darjeeling Gorkha Hill Council (DGHC) in 1988. The second phase of the movement was started in 2007 by Bimal Gurung, Chief of the Gorkha Janmukti Morcha (GJM) party, and it ended with the development of a politico–administrative arrangement named the Gorkhaland Territorial Administration (GTA) in 2011 (Rai 2015; Tamlong 2012).

During the first phase of the movement, S.W. Ladengla, Additional Superintendent of Police joined as the first woman participant and developed an association with the help of elitist Nepali women for the demand of Gorkhaland. After that, GNLF Chief Subhash Ghisingh developed the first women's political wing named Gorkha National Women's Organization in the 1980s. At that time, women at the margins were mobilized to participate in the protests, rallies, campaigning and strikes. However, due to lack of media coverage and violent mode of protest, women were restricted from joining the movement. It also impeded the spread of awareness among the women.

In the second phase, Gorkha Janmukti Morcha's Chief Bimal Gurung developed a renewed demand for Gorkhaland in mobilizing people to support Prashant Tamang (a contestant on the TV show 'Indian Idol'). This led to massive participation of Gorkha women, the reason being that the present movement was based on non-violent mode of protest, democratic in nature and based on Gandhian principles. As a result, a new political women's wing has been formed within the parent body, named the Gorkha Janmukti Nari Morcha (GJNM). After that, the Nari Mukti Morcha vigorously participated in the *Dooar chalo Andolan*, rallies and hunger strikes. They also faced brutal action by the Central Reserve Police Force (CRPF), and two of them died. As a result, the women protested further and went to Jantar Mantar in Delhi on 26 February 2011 (Rai 2015: 85). In this way, the renewed wave of the Gorkhaland movement saw the rise of women's activism in Darjiling. Women got an opportunity to reveal their power, which had also been manifested by their prominent role as the leaders and representatives of the movement. After this, women got a more powerful and respectable position in society, and gender inequality reduced.

Balurghat belongs to the category of small city, with a population size of 151,299 (Census of India 2011). In spite of the location being far away from the Kolkata Metropolitan Area, the city has had a rich history of progressive culture for a long time. The city also has a long history of revolutionary activism since the pre-independence period, such as the Swadeshi Movement against the British Raj and Farmers' Movement against landlords and zamindars, which is commonly known as the Tebhaga movement of Bengal. The Tebhaga movement was actually a movement by the marginal community of share croppers and *bargadars* against the landlords and big land holders, demanding one-third share of crop. However, the educated middle-class people, including men and women from Balurghat, joined hands with these marginal people to fight for equality. The Revolutionary Socialist Party (RSP), the contribution of which is very high in India's freedom struggle, was also formed in Balurghat in 1940. This party with it is revolutionary, and Marxist background had active women members from the very beginning, which paved the way for women's emancipation in the city.

Balurghat is way ahead of many other cities in Bengal in terms of women's emancipation, as the women in this city got the opportunity to educate themselves early, in comparison with other parts. The girls' high school in this city was established in 1926, long before independence. The city got its first college in 1948, and the first women's college in 1970 (Roy 2012). Women in this city always enjoyed better facilities and social atmosphere since the time of independence, which were rare in many cities even in the Kolkata at that time. Besides getting a chance to attend co-educational schools, women used to perform in dramas and other cultural events on public stages alongside men in those days, which was unimaginable in many other cities in Bengal. Nila Kar, a 72-year-old woman of Bardhaman town, who is from Balurghat city and spent her school and college days there, laments, 'The culture in Balurghat was always progressive and was only comparable to the culture of Kolkata city. Women could join men in all kinds of cultural events there, which was unimaginable then in most other cities of Bengal. When I migrated to

Bardhaman for university education in 1964, I had to sacrifice my performances in drama as that was below the dignity of woman in the then Bardhaman city'. If we examine the statement further, we get a picture of Balurghat where women's condition was not as gloomy as it was in many other cities of Bengal.

The analysis of the history of women's emancipation in these two cities leads us to understand why these cities have better capabilities of women councillors. The other two cities do not have any history of women's movements or other kinds of emancipating factors working in the past. This historical background of women's empowerment in the selected study areas proved to be an important factor, as it is highly linked with the present status of capabilities of women discussed in the following chapters.

Appendices

Appendix 3.1: Women's Representation in the Parliament in India

Loksabha	Election year	Percentage of women's representation
1st	1952	4.36
2nd	1957	4.45
3rd	1962	6.27
4th	1967	5.57
5th	1971	4.05
6th	1977	3.5
7th	1980	5.29
8th	1984	7.94
9th	1989	5.48
10th	1991	7.11
11th	1996	7.36
12th	1998	7.91
13th	1999	9.02
14th	2004	8.29
15th	2009	10.86
16th	2014	11.81
17th	2019	14.39

Source: Election Commission of India

Appendix 3.2: The Comparative Representation of Women in Parliament from India and West Bengal: 1952–2019

Loksabha	Election year	Percentage representation of women members to total Loksabha members in India	Percentage representation of women members from West Bengal to total Loksabha members from West Bengal
1st	1952	4.36	6.45
2nd	1957	4.45	9.37
3rd	1962	6.27	6.06
4th	1967	5.57	7.69
5th	1971	4.05	5.26
6th	1977	3.5	4.76
7th	1980	5.29	4.88
8th	1984	7.94	10
9th	1989	5.48	7.32
10th	1991	7.11	7.5
11th	1996	7.36	10.26
12th	1998	7.91	12.82
13th	1999	9.02	11.9
14th	2004	8.29	9.52
15th	2009	10.86	19.05
16th	2014	11.81	30.95
17th	2019	14.39	26.19

Source: www.loksabha.nic.in & www.wikipedia.org.in

Appendix 3.3: Representation of Women in Urban Governance of West Bengal

Districts	Percentage representation of women councillors to total councillors	Observed value (elected seat)	Expected value (reserved seat)	Difference between observed and expected value of representation	Percentage of observed over expected
Bankura	41.38	24	19	5	26.31
Birbhum	38.09	40	35	5	14.28
Bardhaman	38.65	80	68	12	17.64
Coochbehar	46.25	37	26	11	42.30
Hugli	37.62	114	100	14	14
Howrah	36.49	27	24	3	12.5
Jalpaiguri	40.79	31	26	5	19.23
Maldah	40.82	20	16	4	25
Murshidabad	37.32	53	47	6	12.76

(continued)

Districts	Percentage representation of women councillors to total councillors	Observed value (elected seat)	Expected value (reserved seat)	Difference between observed and expected value of representation	Percentage of observed over expected
North 24 Parganas	43.27	254	194	60	30.92
Nadia	37.7	69	60	9	15
Paschim Medinipur	41.05	39	31	8	25.80
Purba Medinipur	37	37	33	4	12.12
Purulia	47.92	23	16	7	43.75
South 24 Parganas	41.18	63	50	13	26
Kolkata MC	38.89	56	48	8	16.66
Uttar Dinajpur	38.96	30	25	5	20
Darjiling	41.98	55	43	12	27.90
Dakshin Dinajpur	32.08	17	17	0	0
West Bengal Average	39.86	56	46	10	

Source: West Bengal State Election Commission 2013–2017

Appendix 3.4: The District-wise Patterns of the Party-wise Representation of EWRs in Cities of West Bengal

Districts	Party-wise percentage representation of elected women				
	AITC	BJP	INC	CPIM	Others
Bankura	48.65	25	100	37.5	12.5
Birbhum	36.36	45.45	27.27	33.33	100
Bardhaman	40.91	100	30.77	36.36	16.67
Coochbehar	71.43	0	57.14	33.33	13.64
Hugli	41.51	50	16.67	36.84	20.69
Howrah	37.93	25	66.67	39.39	0
Jalpaiguri	43.9	50	0.71	90.91	0
Maldah	56	0	50	37.5	33.33
Murshidabad	53.57	36.36	30.51	29.17	39.13
North 24 Parganas	45.09	25	33.33	40.38	33.33
Nadia	38.18		0	35.71	33.33
Paschim Medinipur	44.23	66.67	33.33	22.22	30.77
Purba Medinipur	40	0	0	0	25
Purulia	50	0	38.46	80	42.86

(continued)

	Party-wise percentage representation of elected women				
Districts	AITC	BJP	INC	CPIM	Others
South 24 Parganas	45.19	25	33.33	50	50
Kolkata MC	41.23	0	20	70	12.5
Uttar Dinajpur	48.28	40	37.84	0	0
Darjiling	30.77	100	40	42.86	44.16
Dakshin Dinajpur	21.95	0	0	0	100
West Bengal Average	43.96	30.97	32.42	37.66	31.99

Source: West Bengal State Election Commission 2013–2017

References

Acharya M, Ghimire P (2005) Gender indicators of equality, inclusion and poverty reduction: measuring programme/project effectiveness. Econ Pol Wkly 40(44&45):4719–4728

Avasthi A, Srivastava AK (eds) (2001) Modernity, feminism and women empowerment. Rawat Publications, New Delhi

Banerjee R, Samanta G (2017) Stanio Prashasane Madhyabitta Bangali Narir Absthan. In: Bag KK, Jana RK (eds) Madhyabitto Bangali: Antarmahal. Raktakarabi Publication, Kolkata, pp 222–230

Batliwala S (1994) The meaning of women's empowerment: new concepts from action. In: Sen G, Germin A, Chen LC (eds) Population policies reconsidered: health, employment and rights. Harvard University Press, Boston, pp 127–138

Boserup E (1970) Women's role in economic development. Allen and Unwin, London

Census of India (2011) Retrieved from www.Censusindia.gov.in

Charmes J, Wieringa S (2003) Measuring women's empowerment: an assessment of the gender-related development index and the gender empowerment measure. J Hum Dev 4(3):419–435

Cheston S, Kuhn L (2002) Empowering women through microfinance. Draft Report. UNIFEM. http://www.chs.ubc.ca/srilanka/PDFs/Empowering%20women%20through%20microfinance.pdf

Gaonkar RR (2004) Role of self-help groups in empowerment of women. Paper presented at ISTR Sixth National Conference, Toronto, July 11–14

Garba PK (1999) An endogenous empowerment strategy: a case-study of Nigerian women. Dev Pract 9(1):130–141

Ghatak SK (2010) Political participation of women of West Bengal its nature and extent. Indian J Politic Sci, 71(1) (Jan–Mar), 285–294

Ghosh J (2009) Transcending gender reservation, representation and identity among elected women representatives in municipal politics. Published Master of Science dissertation. School of Interdisciplinary Area Studies, University of Oxford

Hashemi SM, Schuler SR (1993) Defining and studying empowerment of women: a research note from Bangladesh. JSI Working Paper No. 3. John Snow Inc, Washington

IFAD (2000) Tamil Nadu women's development project: completion evaluation. Report 340-IN. IFAD, Rome

Karl M (1995) Women and empowerment, participation and decision making. Zed Books, London

Lahiri-Dutt K, Samanta G, Sil P, Karfa C (2006) Endengering urban environmental management: results from a study of women councillors running a town in India. Women & Environ Int Mag 70/71(Spring):42–46

Madhok S (2003) A limited women's empowerment: politics, the state, and development in Northwest India. Women's Stud Q 31(3/4):154–117

Mazumder V (1989) Reservations for women. Econ Pol Wkly 24(50):2795–2796

Omvedt G (1987) Women and Maharashtra Zilla Parishad elections. Econ Pol Wkly 22(47):1992–1993

Omvedt G (2005) Women in governance in South Asia. Econ Pol Wkly 40(44–45):4746–4752

Oomen MA (2014) Deepening democracy and local governance challenges before Kerala. Econ Pol Wkly XLIX(25):42–46. https://www.epw.in/journal/2014/25/perspectives/deepening-democracy-and-local-governance.pdf. Accessed on 24 June 2014

Rai P (2011) Electoral participation of women in India (key determinants and barriers). Econ Pol Wkly XLVI(3):47–55

Rai A (2015) Women in Gorkhaland movement: a sociological study. Master of Philosophy dissertation, Department of Sociology, School of Social Science, Sikkim University, Gangtok. 14.139.206.50:8080/jspui/bitstream/1/3115/1/Ashlesha%20Rai.pdf Accessed on 28 Aug 2016

Rajagopal S, Mathur K (2000) Women's empowerment through state benevolence. Econ Pol Wkly 35(33):2908–2910

Roy P (2012) A very brief history of Balurghat. http://roypinaki.blogspot.com/2012/11/a-very-brief-history-of-balurghat_7.html Accessed on 20 June 2020

Samanta G (2009) Microfinance and women: gender issues of poverty alleviation and empowerment. Microfinance Rev 1(2009):100–120

Samanta G (2016a) Feminist movements and women's development discourse in India: a special focus on education, health and work. Paper presented at Sorbonne University, Paris, France, 10th February 2016

Samanta G (2016b) Gender and development in India. Paper presented at the Department of Geography, Paris Diderot University, France, 11th March 2016

Samanta G (2016c) Governing small cities in India: a case of women in local governance. Paper presented at the Institute of Geography and Sustainability, University of Lausanne, Switzerland, 5th April 2016

Samanta G (2018) Gender mapping of Burdwan district, West Bengal. Unpublished research project report funded by the University Grants Commission, New Delhi

Sathe D, Klasen S, Priebe J, Biniwale M (2013) Can the female sarpanch deliver? Econ Pol Wkly XLVIII(11):50–57

Sharma K (1998) Transformative politics: dimensions of women's participation in Panchayati Raj. Indian J Gend Stud 5(1):24–28

Sharma S (2015) Creating spaces for women: inclusion through reservation in urban local governance. J Politics & Gov 5(1):29–37

Tamlong DT (2012) Darjeeling diaries. Mani Printing House, Darjeeling

Tapan N (2000) Need for women empowerment. Rawat Publications, Jaipur/New Delhi

Waylen G (1996) Gender in third world politics. Open University Press, Buckingham

Chapter 4
Profile and Background of the Councillors

4.1 Introduction

Research in the social sciences usually depends on the active participation of research participants, and this participation is a very significant component of research. The post-positivist approach contains an element of humanity that connects the people and researcher. It does not consider the people as research objects, but as research participants (Ryan 2006). For this reason, the present chapter focuses on a brief profile of the research participants. The main research participants are the women councillors of the four cities of West Bengal. We have considered all the women councillors (WCs) as research participants, which mean that there is no sampling in choosing the women research participants.

4.2 Profile of the Women Councillors

Darjiling Municipality has 32 councillors, 1 elected from each ward. Among them, 15 are women councillors (WCs). This indicates more than 40% representation of WCs in the municipal board, which is much higher than the reserved 33%. Among the 14 WCs surveyed, 6 are managing higher responsibilities, as they are in the position of Chairman-in-Council (CIC). Among them, one is in the Department of Land, two are from the Urban Poverty and Eradication cell (UPE), and one is in the Waterworks Department. In literature, it is revealed that women representatives are usually placed in the less important departments such as Health and Childcare, as politics is the source of power and income which men attempt to control (Kodoth and Mishra 2011; Rai 2011). However, this view is not applicable to the case of Darjiling Municipality. During the interviews, it was found that there is no proxy

R. Banerjee, G. Samanta, *Negotiating Terrain in Local Governance*, Local and Urban Governance, https://doi.org/10.1007/978-3-030-60663-3_4

among the women in CIC. They are capable of doing the work related to their concerned departments on their own.

The age group of the interviewed councillors ranges from 30 to 50 years. There are three WC who are more than 50 years old (Appendix 4.1). The majority of WCs are married. However, there is one unmarried and one widow among the WCs. Among the WCs, ten are Hindu, while four are from the Buddhist community. This ratio is perfectly in synch with the ratio of the population of the two communities in the city. Of the total WCs, five are from Scheduled Tribe (ST) community, one from Other Backward Classes (OBC), one Scheduled Caste (SC) and seven are from the General category. Most of the WCs who are Buddhist are also from ST background (Appendix 3.1). All the councillors are residents of Darjiling by birth, and only one of them lived in a village near the city previously, which is also under the municipal area.

Balurghat Municipality has 23 councillors, of which 8 are WCs, including the Chairman. This means 34% representation of WCs with respect to the total. Except for the Chairman, there are no WCs in the Municipality's Chairman-in-Council (CIC). Their ages vary from 29 to more than 60 years. If the division of age groups is made, then we observe that there are two WCs within 40 years, one from 40 to 50 years, two from 50 to 60 years and two above 60 years (Appendix 4.2). WC 18 was elected as a councillor at the age of 23 years. There is an interesting fact behind the selection of women candidates to be nominated as a councillor—except for two, all the WCs are either widowed or unmarried. It is also important to mention that the currently married WCs were unmarried when they were elected. Caste-wise distribution shows one from SC and all others from the General category. The habitat of most of the councillors from their birth has been the city. However, two widow WCs have only been here from the past 15 years and have come from Gangarampur and Kolkata.

Raniganj has a total of 22 wards. Among the total, 10 wards have elected WCs. This represents 45% of women in the area's governance. Among the eight interviewed WCs, there is only one CIC councillor in the Department of Health, Childcare and MGNREGS (Mahatma Gandhi National Rural Employment Guarantee Scheme) cell. Others are general councillors. There is a certain pattern of arrangement of power among the WCs, as the WC is fully administered by the Chairman and she tries to control the other WCs (Fig. 4.1).

However, some of them do not want to be under her control. There is one WC from the opposition party elected from ward 9. She is a good leader, but always opposes the work of the Chairman, as well as other councillors. Another WC from the opposition party is elected from ward 21. However, she is discriminated against by the Chairman, and has remained as a namesake councillor. From the questionnaire survey, all these hidden equations about the nature of control and their freedom have been revealed. Most of the WCs are Hindu by religion. However, two are Muslim. Except the WC of ward 1, all of them are married. Caste-wise distribution shows that three are from General category, two Other Backward Castes, one Scheduled Caste and two belong to the Scheduled Tribe community (Appendix 4.3).

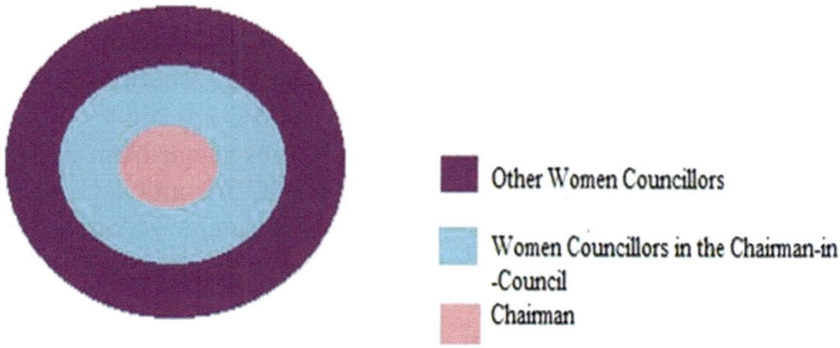

Fig. 4.1 Power hierarchy of councillors. (Source: Prepared by authors on the basis of field knowledge)

In Hugli-Chinsurah, among a total of 30 wards, 11 wards have WCs. However, there are ten reserved seats, and only one WC has been elected in the unreserved seat. It shows 36% representation of women in the governance in relation to total elected seats. Only WC 25 is elected in the unreserved seat.

Among the total 11 councillors, 9 have been interviewed. Their ages vary from 38 to 57 years. There are three WCs within 40 years of age, four WCs from 40 to 50 years and two above 50 years of age (Appendix 4.4). In this context, it is important to mention that women in the middle age group have more experience of life and struggle than the younger ones, as they have first-hand experience of social and political life (Singh 2006). Caste-wise distribution shows that three WCs are from Scheduled Caste, one from Other Backward Caste, and the rest are from General category. All of them are married, and most of them have lived in Hugli-Chinsurah from their birth, securing the identity of a daughter as well as a daughter-in-law in the same city and in some cases, in the same ward.

4.3 Profile of the Men Councillors

In Darjiling, the selected men councillors (MC) are 7 out of the total 16 councillors. Among them, men councillors of ward number 28, 12, 26 and 18 are the Chairman-in-Council (CIC) councillors. The elected persons in wards 12 and 26 are elderly councillors (average age is about 75 years), and they have undertaken responsibilities in the Accounts Committee. MC 18 is in the Mutation Department, and MC 28 is in the Tax and Trade License Department. The age of the other councillors ranges from 40 to 60 years (Appendix 4.5). All of them are married, but MC 21 is a divorcee. Among them, three are from the General category, two from ST, one from Scheduled Caste (SC) and one from Other Backward Caste (OBC). In case of religion, four are Hindu, two are Buddhists, and one is from Muslim community. All of them are residents of that town from their birth.

In Balurghat, among the total men councillors, six have been interviewed. They were elected from the wards 4, 7, 13, 17, 20 and 23. Among them, MC 7 and MC 17 are the CIC councillors. MC 7 has responsibility in the Department of Sanitation and Conservancy, as well as Education and Health. MC 7 is the Vice-Chairman of the Municipality. All of them are married. Their ages range from 39 to 73 years (Appendix 4.6). In case of caste-wise distribution, MC 7 is an OBC candidate, and MC 13 is SC, while the others belong to the General category. Among them, three trace their families originally from Bangladesh; however, they have been residing here since their birth.

In Raniganj, among the total men councillors, six have been selected for this study. They are elected from the wards of 3, 6, 10, 15, 19 and 20. Among them, three councillors are from SC community. Except MC 15, all of them are married (Appendix 4.7). All of them are permanent residents of the city since their birth. Among the selected councillors, there is no MC in the CIC position. However, the Chairman was also selected for the field interview; he is elected from ward 6, but he did not agree to participate in the interview.

In Hugli-Chinsurah, among the total men councillors, eight were selected. However, among the eight MCs, there were two non-responses. The total sample is thus six MCs. The age of the MCs ranges from 45 to 62 years (Appendix 4.8). All of them are from Hindu religion and married. Among them, MC of ward 7 is in a responsible position in CIC.

4.4 Conclusion

From the brief profile of the women councillors of these cities, it is clear that the gender gap in Darjiling is less than that of the other selected cities of West Bengal. In Darjiling, women are elected to the unreserved seats as well, the number of WCs is about 50% of the total councillors, and most of them acquired posts in the Chairman-in-Council (CIC). In other cities, on the other hand, women councillors are mainly elected in reserved seats, and one or two of the total are in CIC position. It is mainly the women of middle age group who have participated for the post of councillors. Caste-wise reservation also exists, and women from Scheduled Caste and Scheduled Tribes are also elected to the reserved seats. A brief profile of the selected men councillors has also been properly explained in this chapter. The men are all married and Hindu, and are mostly from the General category with a few from Scheduled Caste and Other Backward Caste.

Appendices

Appendix 4.1: Brief Profile of the Women Councillors in Darjiling

WC of Wards	Age	Religion	Caste	Marital status
1	46	Hindu	General	Married
2	56	Hindu	ST	Married
5	52	Hindu	General	Married
7	38	Buddhist	ST	Married
10	39	Hindu	General	Married
13	44	Buddhist	ST	Married
14	48	Hindu	General	Married
17	32	Hindu	General	Married
20	32	Hindu	SC	Married
23	40	Hindu	General	Married
27	44	Buddhist	ST	Unmarried
29	41	Hindu	OBC	Married
30	43	Hindu	General	Widow
31	50	Buddhist	ST	Married

Source: Field Survey, 2013–2015

Appendix 4.2: Brief Profile of the Women Councillors in Balurghat

WCs of ward	Age in years	Religion	Caste	Marital status
1	60+	Hindu	General	Unmarried
3	32	Hindu	General	Married
6	63	Hindu	General	Unmarried
9	56	Hindu	General	Widow
15	44	Hindu	SC	Married
18	29+	Hindu	General	Married
21	55	Hindu	General	Widow

Source: Field Survey, 2013–2015

Appendix 4.3: Brief Profile of the Women Councillors in Raniganj

WCs of ward	Age in years	Religion	Caste	Marital status
1	55	Muslim	OBC	Widow
4	32	Hindu	ST	Married
9	45	Muslim	General	Married
12	40	Hindu	SC	Married
13	52	Hindu	General	Married
14	38	Hindu	General	Married
17	43	Hindu	OBC	Married
21	45	Hindu	ST	Married

Source: Field Survey, 2013–2015

Appendix 4.4: Brief Profile of the Women Councillors in Hugli-Chinsurah

WCs of ward	Age in years	Religion	Caste	Marital status
1	38	Hindu	SC	Married
2	42	Hindu	General	Married
9	40	Hindu	General	Married
16	38	Hindu	General	Married
17	43	Hindu	SC	Married
20	47	Hindu	General	Married
23	55	Hindu	SC	Married
25	57	Hindu	General	Married
27	50	Hindu	OBC	Married

Source: Field Survey, 2013–2015

Appendix 4.5: Profile of the Men Councillors in Darjiling

MCs of ward	Age in years	Religion	Caste	Marital status
9	46	Hindu	General	Married
12	75	Hindu	General	Married
18	44	Muslim	OBC	Unmarried
21	38	Buddhist	General	Married

(continued)

MCs of ward	Age in years	Religion	Caste	Marital status
25	39	Hindu	General	Married
26	70	Hindu	SC	Married
28	59	Buddhist	ST	Married

Source: Field Survey, 2013–2015

Appendix 4.6: Profile of the Men Councillors in Balurghat

MCs of ward	Age in years	Religion	Caste	Marital status
4	32	Hindu	General	Married
7	48	Hindu	General	Married
12	59	Hindu	OBC	Married
13	53	Hindu	SC	Married
17	73	Hindu	General	Married
23	51	Hindu	General	Married

Source: Field Survey, 2013–2015

Appendix 4.7: Profile of the Men Councillors in Raniganj

MCs of ward	Age in years	Religion	Caste	Marital status
3	46	Hindu	SC	Married
10	59	Hindu	SC	Married
15	55	Hindu	General	Unmarried
19	39	Hindu	SC	Married
20	49	Hindu	General	Married

Source: Field Survey, 2013–2015

Appendix 4.8: Profile of the Men Councillors in Hugli-Chinsurah

MCs of ward	Age in years	Religion	Caste	Marital status
3	40	Hindu	General	Married
7	45	Hindu	General	Married
10	49	Hindu	General	Married
14	62	Hindu	General	Married

(continued)

MCs of ward	Age in years	Religion	Caste	Marital status
18	53	Hindu	General	Married
22	54	Hindu	General	Married
26	48	Hindu	General	Married

Source: Field Survey, 2013–2015

References

Kodoth P, Mishra US (2011) Gender equality in local governance in Kerala. Econ Pol Wkly XLVI(38):36–43. https://www.epw.in/journal/2011/38/insight/gender-equality-local-governance-kerala.pdf Accessed on 02 July 2012

Rai P (2011) Electoral participation of women in India (key determinants and barriers). Econ Pol Wkly XLVI(3):47–55

Ryan AB (2006) Post-positivist approaches to research in researching and writing your thesis: a guide for post graduate students. MACE: Maynooth Adult and Community Education, Maynooth, pp 12–26. eprints.maynoothuniversity.ie/874/. Accessed on 7 Apr 2018

Singh SK (2006) Women empowerment in India: emerging leadership in urban U.P. In: Singh UB (ed) Empowerment of women in urban administration (experiences and strategies). Serials Publications, New Delhi, pp 61–64

Chapter 5
Work, Leisure and Quality of Life of Women Councillors

Having, Loving, and Being are catchwords for central necessary conditions of human development and existence. It is clearly assumed that there are both material and non-material basic human needs, and that both types of need have to be considered in indicator systems designed to gauge the actual level of welfare in a society.—Erik Allardt 1993

5.1 Introduction

The socio-cultural life of the women councillors changes with their participation in politics. Their work has shifted from daily household duties to large-scale management of their wards. In this chapter, the changing scenario of their socio-cultural life has been analyzed using the framework of having, loving, and being principles.

The concept of 'Having, Loving, Being: An Alternative to the Swedish Model of Welfare' was given by Erik Allardt in the book *The Quality of Life*, edited by Martha C. Nussbaum and Amartya Sen (1993). This concept is based on three main features: having, loving, and being. Here, the concept 'having' is used to see what they have and their perception about their having, after becoming a councillor. 'Loving' signifies their loving persons, and the influence of those persons on their work and on the decision-making abilities of the women councillors. 'Being' stands for the extent to which they participate in decision-making, political activities and opportunities for leisure-time activities, enjoyment and meaningful activities.

In the present research, a few indicators have been added on the basis of the priority of the women councillors (WCs). For the principle of having, WCs' perception about salary, funds allotted to them for conducting their work, stress related to work and years of formal education, distance or hurdles of the journey has been

R. Banerjee, G. Samanta, *Negotiating Terrain in Local Governance*, Local and Urban Governance, https://doi.org/10.1007/978-3-030-60663-3_5

considered. In case of loving principle, the attachments and contacts of WCs with family, colleagues, local community, and other social organizations have been analyzed. The indicators of being for the WCs are the same as stated.

For conducting this survey, a questionnaire based on the variables of having, loving and being principle has been set. Having, in the context of women councillors, signifies their perception about their honorarium, the hurdles on their journey—for instance, distance from the Municipal office, modes of transport, and the cost of travel—and, lastly, their stress related to work. Loving signifies the relationship of the women councillors with the close people in their lives. Being is related to leisure time and how they spend it, the opportunities they aspire to from their position, and the incentives such as honour, awards, and prizes. It signifies the role they perform as women councillors. Besides their work, it also relates to how they enjoy their leisure time and the quality of their life, after being elected as women councillors. On the basis of those criteria, the changing socio-cultural lives of the women councillors have been analyzed besides their changing time-use pattern. From the field survey, it has also been found that the pattern of changing socio-cultural lives in respect of having, loving and being principles also changes with space specificities of the cities.

5.2 Quality of Life of Women Councillors

Quality of life is a crucial aspect of the lives of women in general, as they are often overburdened with the double pressure of working outside and managing all the domestic affairs. Quality of life depends on more than just material well-being. It is closely linked to emotional components, and is difficult to measure with quantitative data. Subjectivity is endemic in the measurement of quality of life. Following the work of Erik Allardt, we have tried to explore the quality of lives of the women councillors through having, loving, and being parameters. The methods followed for this section are predominantly qualitative with informal discussions. The discussions were carried out through repeated interviews with the councillors individually. Within the 'having' principle, we have tried to capture their gains and pains as a councillor. The 'loving' principle is focused on their emotional relation with family and friends. Leisure is also an important component of quality of life, and this has been understood with the 'being' principle.

5.2.1 Case of Darjiling

5.2.1.1 Having

From the field work, it is found that the women councillors of Darjiling are highly dissatisfied with the salary and the fund allotted to them to do their work. They express their perception that the salary is very low, at INR 1575 per month for general councillors, and INR 2200 per month for Chairman-in-Council (CIC) councillor. They further face the compulsion of spending a portion of their salary by giving *chanda* (token money) to the club during festivals. WC 17 (women councillor elected from ward 17) argued that the low salary is an indirect encouragement to practice corruption. This argument is similar to the statement by Ghosh and Lama-Rewal (2005) in their study. They stated that the government spends INR 80,000 per month for a single Member of the Legislative Assembly (MLA) for his or her personal assistant, car, cell phone, etc. Such facilities should be extended to the councillor as well, because they do maximum work. WC 29 also added that the salary is very low in comparison with the recent increase in price of everyday commodities. The WCs from wards 1 and 2 said that they travel long distances and spend INR 40 daily as transport cost to go to the Municipal office. They also spend some money on giving *chanda* very frequently. For that reason, their monthly salary is finished within 2 to 3 days if they come to the Municipal office daily. The WC had suggested that the salary should be increased up to INR 2000–3000 per month, but this was ignored by the officials at that time. The increase in the salary of the councillors and increase in the availability of funds through the legal system is also suggested by Drage (2001).

In case of the fund, councillors blame the present state government for not releasing sufficient funds for Darjiling. The Bengali-speaking CIC has also mentioned that the government releases the fund to hilly regions, keeping in mind the condition and expenditure of the plain regions. As a result, the fund is very low and insufficient for the hills, due to the high cost of transport here. According to her, the value of INR 800 for the plains is the same as INR 8000 in the hills. Similarly, the government gives cotton garments to poor people of the hills, which is unsuitable for the cold, hilly climate.

Though they have low salaries, the WCs perform highly responsible work. For that reason, they sometimes feel stressed out. WC 23 has reported an interesting fact that she does meditation to reduce the stress of her work. WC 20 said, ‘When I am not able to give time to my home due to pressure of office work, I am in stress’. WC 27 said that the frequently placed multiple demands of the local people are the main source of her stress. WC 2 said that she cannot take care of herself and feels totally lost due to her excessive work pressure. WC 29 stated that lack of privacy is a cause of stress for her. She has described a story related to it: One day, her sister-in-law invited her on an occasion. She was also ready to go there. However, due to an urgent call for a meeting, she had to cancel the invitation. WC 30 said that the irresponsible nature of the office staff is a cause of stress for her.

Distance is a major hurdle in the journey of the women councillors of Darjiling. Due to distance and terrain conditions, the cost of travel is high. Most of the women councillors do not come to office daily due to the distant location of their house from the Municipal office. For example, the WC from ward 2 comes every day from her distant house. However, the other women councillors of her adjacent wards like WC from ward 1 do not come daily and also encourage WC 2 not to come to office daily.

5.2.1.2 Loving

While analyzing the variables of the loving aspect, all the women councillors happily talked about their loving personal relationship with family, friends and other councillors, and with the people of their locality.

Most of the women councillors mentioned the name of their husband and son as loving persons. Women councillors of older age mentioned the name of their son-in-law and younger members of their family as loving persons. Their husband and other loving persons have provided opinions and suggestions to their work, and also helped them in household chores.

All of them have an active friendship circle. Only three of the women councillors within 40 years of age use social media such as Facebook and other social networking sites to stay in contact with their friends. The councillors also mentioned the name of another councillor as being close to them. The reasons behind their closeness are the similarities in ideology and the location of their house in the same neighbourhood. For example, WC from ward no 1 mentions the name of WC 2, as both of them go to the Municipality together. WC 23 had a different perception of friendship with other councillors. She has no closeness with the other councillors. She explained that this was because she has come here to do her work, and she does only that.

5.2.1.3 Being

Being is related to leisure time, how they spend it, and the opportunities they aspire for, from their position and other incentives such as honours, awards, and prizes. All the WCs spend their leisure time with their family, and enjoy relaxing by watching TV, reading the newspaper, and so on. However, two of them mentioned that they do yoga in their leisure time. On the question of aspiration for opportunities, a few of them talked about the opportunities for their personal betterment. Their main priority is to get transport facilities to reach the office, and the other is to get encouragement from the officials for their performance. Being ignored by officials sometimes suppresses the confidence of the women leaders morally as well as psychologically, which is also observed by Rizvi (2006).

5.2.2 Case of Balurghat

5.2.2.1 Having

Women councillors of Balurghat are highly dissatisfied with their salary. However, they are satisfied with the fund allotted to them. Most of them get a salary of ₹1500 per month. The Chairman spends all her salary to enrich her party fund so that they can work for social development. The honorarium is helpful to generate a little income for the housewives. However, all of them acknowledge the honour they get from their position.

In case of fund availability, except for two women councillors, all of them are dissatisfied. In this context, WC 17 has said that she wants to establish 30 houses in the slum area; however, due to inadequate funds, it remains unfinished.

Most of the women councillors never feel stress due to work. One reason for this is that the Chairman always handles the stressful situations and helps to bring confidence among them. The other reason is that most of them are unmarried, that is why they have no additional pressure of housework.

Distance and hurdles of the journey are not a problem of the women councillors of Balurghat. The reasons are (a) the city is situated in the plains, and (b) the size of the city is small. On the meeting days and on days of excessive work, the Chairman provides her own car to drop the women councillors to their house.

5.2.2.2 Loving

In Balurghat, most of the widowed women councillors mention the name of their son or daughter as loving persons. Though WC 21 is a widow, she acknowledges the name of her husband. Women councillors of older age, as well as the unmarried WCs, mentioned the name of their mother as a loving person. These loving persons provide opinions, encouragement and suggestions to them to do their work better, and also help them in household chores.

All of them have an active friendship circle within the sphere of councillors. They respect the suggestions which are given by the woman Chairperson. WC 21 and WC 17 have close relations with each other, as they go to the Municipal office together on a daily basis. However, all of them are close to each other, and have a sense of unity. Officials are also very helpful to them. For example, WC 10 has mentioned the name of a Municipality staff member who helped her to learn how to write official letters. WC 7 also acknowledges the name of her ward development secretary for his generous help.

5.2.2.3 Being

All the women councillors spend their leisure time with their family, and enjoy their leisure time by watching TV, reading the newspaper and so on. However, WC 1 and 10 spend their leisure time in doing social work. WCs are happy with the available support they receive from the Chairman as well as from the other officials.

In case of incentives, all the councillors are pleased because of the honour they have got from their position. The Chairman also gives moral encouragement to them.

5.2.3 *Case of Raniganj*

5.2.3.1 Having

The perception about the salary among the women councillors depends on their overall economic status. For example, WC 13, 14 and 9 have no problem with the salary, and they give their salary to the party fund. However, WC 9 raised her voice on behalf of other councillors to increase the amount of honorarium. As a result, it was raised from ₹700 to ₹1000. WC 17 said that the salary is very low in respect of expenses for mobile phones and transport. She also added, 'If we come to Municipal office daily then the salary will be insignificant in terms of expenses'. In a similar study, Ghosh and Lama-Rewal (2005) argued that the function of a councillor entails expenses on telephone, transport, and tea, which are not covered by his or her monthly allowances.

Most of the women councillors are not satisfied with the fund available for operating different government schemes. All of them have stress related to their work.

In the plains, though there is less hurdle of journey, the distances and cost of transport are matters of concern for the women councillors. For example, WC 1 has to travel a very long distance, and cycle rickshaw is the only mode of transport to reach there. As a result, she has to spend ₹40 to reach the Municipal office. The wards 4 and 21 are also situated in a distant location.

5.2.3.2 Loving

Husband, daughter and son are the loving persons for most of the women councillors. They usually discuss everything with their husbands while taking decisions related to their work.

The women councillors of Raniganj reveal a very different picture than the councillors of other places. They do not have any active friends' circle. They have no account on any social networking sites, as they are not habituated to it.

Most of them took the name of WC 13 as their close friend in the councillors' circle. The reason is that she is good in conversation with all of them. WC 1 said that she has a very good relationship with WC 9 (councillor from the opposition party, All India Trinamool Congress). On the other hand, WC 9 stated that she has no close friends in the councillors' circle because of mismatch in thinking, and she maintains only formal relationships with them.

5.2.3.3 Being

Most of the women councillors of Raniganj spend their leisure time by reading books. Except for reading books, WC 1 spends her time with her little grandson and granddaughters. WC 9 spends her leisure time in performing *Namaz* (worshipping God).

In the case of incentives and opportunities, different councillors have different viewpoints. For example, WC 13 and 14 do not opt for any opportunities. They state that they are happy with whatever they have from their position. WC 12 stated that she does not need anything except respect from the people. WC 1, 4 and 21 have opted for transport facilities as incentive, as their houses are located at a distance from the Municipal office.

5.2.4 Case of Hugli-Chinsurah

5.2.4.1 Having

In case of perception of salary, all the WCs stated that they get ₹1500 per month as their salary, and ₹200 for mobile recharge. All of them perceive the salary is low. They are housewives, and financially dependent on their husbands' income. They have to do social work and bear transport costs on their own. That is why, besides their low honorarium, they take money from their husband. In this context, WC 20 says, 'We are in the lower position than the Municipality labour in terms of salary. We have no respect in the Municipality'.

This argument focuses on the official status of councillors. It also signifies that local politics is a part-time, voluntary activity as proven by nominal nature of councillors' allowances. This is also observed by Ghosh and Lama-Rewal (2005) in their study.

In case of fund availability, WCs are not satisfied. They have said, 'We have lot of work to do. Due to non-availability of the fund, the works are not yet sanctioned'. WCs also feel stressed as there is no fixed time when the people come to their house for their work. Sometimes, they feel disturbed because of that. Women have additional responsibilities, and a different environment in their houses, but the councillors' work has no fixed time duration, and people approach them at all hours. That is why WCs sometimes feel disturbed.

The distance and hurdles are comparatively less among the WCs of Hugli-Chinsurah, as the Municipal office is well-connected with all modes of transportation. The WCs go to the office once a month during the board of councillors (BOC) meeting. For other work, their ward secretary, party members, and their husbands come to the Municipality on their behalf.

5.2.4.2 Loving

The WCs stated the name of their husband as their loving person. Their husbands help to take proper decisions in case of ward development. Husbands also perform the outdoor works of the councillors, such as ward visits, and work related to the Municipal office. WCs also acknowledge that they have received a certain amount of popularity because of their husband.

The WCs have no friendship circle, and they are not habituated to use the popular social networking media such as Facebook, Twitter, etc. However, they have developed closeness with the other WCs on the basis of their location in two contiguous wards. For example, WC 1 has very close relations with WC 2 because their wards are situated nearby. Other WCs maintain a political relation among themselves. Such division among elected women under the banner of political parties does not help to unite them for the causes affecting women, which is also observed by Rizvi (2006) in his study.

5.2.4.3 Being

WCs spend their leisure time in their own way. For example, WC 16 practices dance during her leisure, WC 25 does stitching work, and WC 23 practices singing. However, most of the WCs say that they give time to their family, or enjoy watching TV or reading books. Only WC 17 said that she has no time for leisure. She does not even have the time to read *Rabibasariya* (a special section of Bengali newspaper for Sunday) even on Sundays.

There are no incentives and opportunities for the WCs. However, they have demanded several incentives, such as an assistant, computer and transport facilities. In this context, WC 23 has a different demand. In her words, ‘I do not need an assistant as I have my husband who can help me. However, we (she and her husband) have given services as councillors. If Municipality provides health insurance to us, it will be very helpful for us during medical emergency’.

Similarly, WC 25 has demanded for pension. These two arguments from the councillors reveal that they need security for their future from their work as councillors. According to them, since they provide services to the citizens during their tenure, therefore it is the responsibility of the government to provide security for them in terms of assuring health insurance and pension.

5.2.5 *Overall Comparison of the Cases*

From the previous discussion of having, loving, and being principles of the women councillors, it is clear that the honorarium and fund availability are limited in all the cities of West Bengal. WCs face stress during their work, but the main hurdles of their functioning are distance, mainly in the case of Darjiling. In other cities, the hurdle of journey is the cost of transport. They want to get honour, awards, and prizes for their work. They think that it will increase their enthusiasm for their work. They also want to have more reward of their duties in the formal structure of the municipal system through pensions and insurance.

They have a feeling of fulfilling a part-time duty because of the low honorarium and no added facilities. They work for the people in addition to attempting to fulfil various needs of their ward. Their continuous engagement with the work also deserves a proper salary to allow them to undertake their work wisely; otherwise, they will lose interest in the long run. How far this changing socio-cultural life of the women councillors of the cities affects their time-use pattern is discussed in the next section of the chapter for better understanding of their lifestyle in the role of councillor.

5.3 Time-Use Survey[1]

Women's participation in local governance leads to additional investment of time for work as a local representative in their daily routine. This in turn changes their social life. In local governance, women have to undertake quantum of work, similar to the amount of work taken up by a male representative. Lahiri-Dutt and Sil (2014) stated that in the house, gender attitudes, gender roles and others' expectations result in uneven distribution of work between men and women. Therefore, women engaged in earning for the household, or those who have an engagement in the public sphere tend to do more housework when compared to a man. A similar argument has been given by Bittman and Wajcman (1999) in another context, where they stated that norms of behaviour and masculinity prevent men from doing housework, that is, the more a husband depends on his wife, less he works at home. For these above-mentioned reasons, the lack of time to manage both household affairs and the responsibilities of a councillor is a major challenge for a majority of women in urban governance. Ghosh and Lama-Rewal (2005) also stated that sometimes the women councillors feel guilty about remaining out of their homes for long hours. This is a psychological problem faced by most of the women councillors. This kind of feeling is the result of internalization of patriarchal norms by women themselves.

Women councillors are required to spend a certain amount of their time in the Municipal office, and in their concerned wards. However, as women, they also have to allocate a certain fixed amount of time daily for their routine household chores and their own maintenance, especially in the case of married women. The household chores are divided into household-related work and activities performed within the home, and household-related work and activities performed outside. There are two types of household activities which are performed outside the home—regular and occasional work. Regular household work performed outside the home includes buying groceries, dropping children to school and coaching classes, and attending cultural activities. Occasional household work performed outside the home includes shopping for household items including clothes for children, visiting the doctor for medical treatment of family members, collecting groceries from ration shops, carrying out banking transactions, and paying telephone and electricity bills. Thus, household duties have actually expanded and diversified. A woman has to allocate

[1] Part of this section was published in Banerjee, Riya (2017). 'Does Time matter?—A Study of Participation of Women in Urban Governance' *Space and Culture, India*, ACCB Publishing, 4(3): 62–73.

her time between these three competing types of work—work of a local representative, household work, and time for her own leisure. For these reasons, the present chapter focuses on the changing nature of time use patterns of women in light of their political participation in urban governance in the cities of West Bengal.

To identify (a) the hours spent on work by the women as local councillors; (b) the time they spend in their homes to perform household work and the influence of it on their work as a local councillor; and (c) the exact time of their leisure and how they spend it, considering the influences on household work as well as the work of a councillor, the time-use diary is used.

This diary is used to enable respondents to report all activities undertaken over a prescribed period of time, noting the time when each activity begins and ends. To collect the required data, a 24-hour diary has been used, which was subdivided into two basic types—a full-time diary and a light time diary. The full-time diary is again of two types: open time interval and fixed time interval. In this study, the open time interval full-time diary is used to retrieve information on how the WCs spend their time throughout the course of a day. In this diary, the respondents reported what activity they were doing when they began the day, what activity came next and at what time this activity began and ended, and so on, successively through the 24 hours of the day. The interval of time within which an activity is reported is not fixed. Such recording of the duration of each activity is called an episode (Hirway 1999; Budlender 2007; Department of Economic and Social Affairs S.D. 2005; European Communities 2004; The Community GPI Survey Part – II n.d.; Task Force on Time Use Surveys 2013). Time is measured in hours and minutes per day, out of a total of 7 days in 1 week (weekly average), which was our reference period for collecting the data. Recording of activities is targeted for a normal day, not for a weekly variant or abnormal day. Rajivan (1999) clarifies the concept of weekly variant day. This day allows for a general day off, and the abnormal day is a day in which all usual functions of the household may be altered for a special reason, such as due to the arrival of an unexpected guest, an illness or even a local festival.

In the present context, the 1995 classification of activities recommended by the United Nation Development Programme (UNDP) has been followed (UNDP 1995). The trial international classification for time-use activities differentiates activities on the basis of whether an activity is within the System of National Accounts (SNA) production boundary, or within the general production boundary outside the SNA boundary or non-productive. In total, 154 activities have been grouped into a ninefold classification, and these nine groups of major activities have been further classified into three broad divisions by the United Nations: (a) SNA activities [(I) primary, (II) secondary and (III) tertiary activities which are paid works]; (b) the extended-SNA activities [(IV) household maintenance, management, and shopping for own household; (V) care for children, the sick, elderly, and disabled for own household; and (VI) community services and help to other households]; and (c) non-SNA activities [(VII) learning; (VIII) social and cultural activities, mass media; and (IX) personal care and self-maintenance].

5.4 Time-Use of the Women Councillors

A detailed time-use survey of the WCs in the four cities of West Bengal provides a comparative picture of their contribution on three types of time-use activities: System of National Accounts (SNA), extended-SNA, and non-SNA. In this context, the SNA, extended-SNA, and non-SNA activities (already discussed in detail) have been analyzed on the basis of their employment status—that is, employed (those who are engaged in paid employment besides their work of local councillor), and unemployed—and marital status (married, unmarried and widowed) (Table 5.1). Here, the total time invested by the WCs in their work as the local representative, and in other economic activities (paid employment) is considered as the WC's time for doing SNA activities.

5.4.1 Comparison Between Married and Unmarried Unemployed Women Councillors

To compare the contribution of WCs in terms of the weekly average time spent on SNA, extended-SNA and non-SNA activities, the data has been analyzed on the basis of their employment as well as their marital status.

Table 5.2 clearly depicts the greater workload of household duties of the unemployed WCs in the selected cities of West Bengal. However, Ghosh and Lama-Rewal (2005; quoted in Ghosh 2009) stated that unemployed WCs can dedicate more time to the work of the local councillor as they have no extra burden of paid employment like the men have. The findings of this study thus remain contrary to the findings as reported in the literature. It is quite obvious that the married unemployed WCs spend more time in extended-SNA activities (average 66–84 hours in a week) than the widowed and unmarried unemployed WCs.

Keall et al. (2007) stated in this context that unpaid care and household chores are significant functions necessary to maintain the standard of living and well-being of the family in a society. However, such work may limit or prevent people from participating in paid work. This in turns affects their incomes and access to goods and services. On the contrary, it is observed in Darjiling that married unemployed WCs spend less time in extended-SNA activities than the married unemployed WCs of other cities of West Bengal as there is less gender division of work between men and women.

Except for the married unemployed WCs of Darjiling and Balurghat, all the WCs of the other two cities get less time to spend in SNA activities (on average, 33–43 hours in a week) due to their involvement in extended-SNA activities. In Balurghat, the discipline of the woman Chairperson keeps the WCs active in their SNA activities (work of the local councillor) despite their household duties. For this, the married, unmarried, and widowed unemployed WCs spend more or less the same time for their SNA activities (on average 36–44 hours in a week). In

Table 5.1 Profile of the women councillors on the basis of marital and occupational status

Cities	No of total councillors	Total women councillors	Surveyed women councillors	Marital status			Occupational status	
				Married	Unmarried	Widow	Employed	Unemployed
Darjiling	32	15	14	12	01	01	04	10
Balurghat	23	08	07	3	2	2	03	04
Raniganj	22	10	08	07	–	01	–	08
Hugli-Chinsurah	30	11	09	09	–	–	–	09

Source: Field Survey, 2013–2015

Table 5.2 Time-use of unemployed women councillors

	Time-use of unemployed women councillors with different marital status				
	Weekly average time in hours				
Cities	Marital status	Hours of SNA activities	Hours of extended-SNA activities	Hours of non-SNA activities	Total hours
Darjiling	Married	45.18	66.69	56.13	168
	Unmarried	36.92	74	57	167.92
	Widow	49	70	49	168
Balurghat	Married	43.75	71.75	52.5	168
	Unmarried	42	63	63	168
	Widow	36.19	82.83	49	168.02
Raniganj	Married	42.5	75.14	50.29	167.93
	Widow	59.5	52.5	56	168
Hugli-Chinsurah	Married	33.22	83.67	51.11	168

Source: Field Survey, 2013–2015
Note: The figure of total time for each may not be exactly equal to168 due to rounding off effect

Hugli-Chinsurah, on the other hand, married unemployed WCs spend less time in SNA activities (on average 33 hours in a week). They only attend the Municipal office, while the other work of the local councillor is done by their associated male members (husband or party members). They are thus elected in proxy seats. This might have significant implications, as Hossain (2012) observed that the women have nothing to do for the actual well-being of the community, as their work is done by their husbands.

From Table 5.3, it is clear that the unemployed WCs of Hugli-Chinsurah spend less time in the Municipal office and ward visits (on average, 3–6 hours in a week). The reason is the men of the family and of the political party do the outdoor work on behalf of the women councillors. In Raniganj, five out of eight WCs undertake ward visits for 3–7 hours in a day, while the others prefer to wait in the local party office of their ward to meet the common people (about 16–28 hours in a week). Hanson and Pratt (1995) stated in this context that workspace choices in formal settings outside their home are also gendered, as their work is mainly concentrated at home. The selection of work space for the women depends on the distance from their houses and opportunities to access means of transport, jobs with less spatial mobility, and so on.

The unemployed WCs of Balurghat and Darjiling spend more time on their work as the local councillor outside their home (i.e. for ward visits and municipality work). Among them, the unmarried and widowed WCs contribute more time in work outside their homes (Table 5.3) because they have less burden of household chores. In Darjiling, unemployed WCs prefer to stay in the Municipal office (on average, 22–35 hours in a week) to get to know the problems of the common people, and to get essential information regarding the development of their ward. Here, the

Table. 5.3 SNA activities of unemployed women councillors

Cities	Marital status	Distribution of SNA activities of unemployed women councillors of different marital status			
		Weekly average time in hours			
		Hours spent in municipality	Hours spent in ward visit	Hours to meet the local people at house or at party office	Total hours of SNA activities
Darjiling	Married	30	15.18	–	45.18
	Unmarried	22.92	14	–	66.92
	Widow	35	14	–	49
Balurghat	Married	33.25	10.5	–	43.75
	Unmarried	31.5	10.5	–	42
	Widow	29.19	7	–	36.19
Raniganj	Married	12.78	13.71	16	42.49
	Widow	31.5		28	59.5
Hugli-Chinsurah	Married	5.99	3.5	23.72	33.21

Source: Field Survey, 2013–2015

ward visit is quite hard for the WCs because of hilly terrain and uneven distribution of wards, along the different spurs of the hill.

In case of the time spent on non-SNA activities, unemployed married WCs of Darjiling spend more time (on average 56 hours in a week) than WCs of the other cities of West Bengal (Table 5.2). The reason for this, as already stated earlier, is that there is less gender division of work between men and women, so WCs spend less time in household activities. Schor (1991, quoted in Bittman and Wajcman 1999) stated that the scarcity of time and the paucity of leisure time are at the centre of the discussion about the quality of contemporary life. Moreover, for the unmarried WCs in all cities, a reduction in time spent on extended-SNA activities results in a corresponding or greater increase in time spent on non-SNA activities (on average 57–63 hours in a week).

5.4.2 Comparison Between Married and Unmarried Employed Women Councillors

In this section, we have made a comparison between married and unmarried employed WCs within Darjiling and Balurghat. This is because there are no employed WCs in Raniganj and Hugli-Chinsurah. In Darjeeling, married employed WCs spend less time (on average 59.5 hours in a week) in extended-SNA activities than the unemployed married women councillors (on average 66.69 hours in a week) (Tables 5.2 and 5.4).

Table 5.4 Time-use of employed women councillors of Darjiling and Balurghat

Cities	Marital status of the women councillors	Weekly average time in hours			
		Hours of SNA activities	Hours of extended-SNA activities	Hours of non-SNA activities	Total hours
Darjiling	Married	57.75	59.5	50.75	168
Balurghat	Married	70	42	56	168
	Unmarried	98	14	56	168
	Widow	73.5	31.5	63	168

Source: Field Survey, 2013–2015

The reason is that the employed WCs need to share their time between other economic activities and their work as the local councillor. As a result, they get less chance to spend their time in extended-SNA activities. Most of the WCs are employed as teachers in English-medium model schools (morning section). However, they spend less time (on average 17.5 hours in a week) for other economic activities than the time they spend in the Municipal office (on average 26.25 hours in a week) (Table 5.5).

The reason is that their working hours start very early in the morning, at about 5 am, and end at 3.30 pm. They cannot work up to 5 pm or more because of cold climatic conditions. As a result, in Darjiling, hours of work for the local councillors start from 12 noon to 3 pm, and there is no overlap in sharing the time between other economic activities, and the work of the local councillor. In Balurghat, most of the employed WCs are also teachers. The unmarried WCs in Balurghat spend more time in SNA activities (on average 98 hours in a week), mostly on municipality work (on average 49 hours in a week) (Tables 5.4 and 5.5). They spend less time on household chores as they have fewer burdens of household duties as compared to the married unemployed WCs (Banerjee 2017).

Table 5.4 also illustrates the fact that the widowed employed WCs of Balurghat spend more time in non-SNA activities, as the burden of household duties is performed by their daughter(s)-in-law and sons. They spend their leisure with their grandsons and granddaughters. In Darjiling, employed WCs spend more time in non-SNA activities (on average 57.75 hours in a week).

In order to undertake the responsibilities of the councillor, the additional investment of time for the work of the local councillor is also related to the honorarium paid to them, which is discussed in the next section.

5.4.3 Time-Use Activities and Its Relation with Honorarium

Rajivan (1999) stated that time is a direct source of utility. According to Harpreet and Anupama (2014), in today's world, the importance of anything is measured in terms of money; the time and worth of work are judged from the payment made for

Table 5.5 SNA activities of employed women councillors of Darjiling and Balurghat

Cities	Marital status	Weekly average time in hours			
		Hours spend in municipality	Hours spend in ward visit	Hours of other economic activities	Total hours of SNA activities
Darjiling	Married	26.25	14	17.5	57.75
Balurghat	Married	28	14	28	70
	Unmarried	49	21	28	98
	Widow	31.5	14	28	73.5

Source: Field Survey, 2013–2015

it. The distribution of time for paid and unpaid work is highly gendered in terms of men and of women. The domestic work is undervalued and unprotected. Pandey (2000) stated that there is no availability of state-level data regarding the time spent by women on domestic activities. From the analysis of the activities of the women councillors, it is observed that among the unemployed WCs, married councillors spend an average of 31 hours, while unmarried and widowed councillors spend 39 hours a week for their work as local councillors (i.e. municipality work and ward visits). In addition, employed married WCs in the cities spend an average of 41 hours a week for their work as local councillors. Unmarried employed WCs of Balurghat spend about 70 hours a week for their work as the local councillors.

In Darjiling, the honorarium of the WCs ranges from INR 1575 to INR 2200 per month. It is also dependent on their position—whether they are in the Chairman-in-Council or not. In Balurghat, the honorarium is INR 1500 per month. In Raniganj, it was raised from INR 700 to INR 1000 per month with an additional INR 200 for mobile recharge per month. In Hugli-Chinsurah, WCs receive an honorarium of INR 1500, with an additional INR 200 for mobile recharges.

Except for the WCs who are employed, the honorarium is problematic, keeping in mind the present expenditures of the WCs. It is quite clear that the perception about the honorarium among the WCs depends on their status of income as well as expenditure. WCs from lower-income group families raised their voices to increase the amount of honorarium. In all the cities, the honorarium of the councillors varies around INR 1500. In comparison with the hours of work of the women councillors, the honorarium is very low, which is stated by most of the WCs interviewed in all the cities.

5.5 Conclusion

From the present study, it is evident that the time-use pattern of the WCs in the cities differs widely. For example, unemployed WCs with different marital status spend more time in extended-SNA activities, which are mostly related to their household work. Among the SNA activities, they spend more of their time at the Municipal office, or at the local party office than in ward visits. Employed WCs spend more time in SNA activities, as they have to undertake the work of a councillor as well as other economic activities. The gap between the hours spent by the WCs on SNA, extended-SNA and non-SNA activities is very low in the case of Darjiling. However, the gap is very high in case of Balurghat, as they spend more time in the Municipal office, irrespective of their other economic activities. Another reason is that the time spent on the other economic work by the WCs does not coincide with the time they spend in the Municipal office. In Darjiling, married, unemployed WCs spend less time in extended-SNA activities than the unmarried and widowed WCs. However, in other cities, they spend more time in extended-SNA activities.

The time-use pattern of the local councillors' work shifts from Municipality work and ward visits to work from home in the north and south of West Bengal. It is found that WCs of northern Bengal are less bound to their household activities than those of south Bengal. As a result, WCs of south Bengal spend relatively less time in the Municipal office as well as in ward visits. This is also a cause of their proxy representation in local governance. Such proxy representation is further encouraged by the limited economic opportunities in this field. Therefore, though household work is considered as extended-SNA activity, it still remains unpaid in a developing country like India for emotional reasons. Thus, household work often hinders the path of the women to work in the field of local governance. For instance, most political parties are hesitant to include women in their core group, as they are required to fulfil their household obligations as well (Panday 2008; Haug 2012).

The work of the councillors is also a time-consuming job, as it is related to service delivery and problem-solving. However, it is wrapped in the cover of social service and therefore, remains underpaid. The honorarium is very low, only INR1500–2200 per month, and also depends on their position—whether they are general councillors or Chairman-in-Council. Such lack of economic freedom and the enormous stress on the use of time is a major cause of discouragement for women to enter this field. Such initial discouragement does not offer a bright vision of the future, despite the reservation for women in urban governance.

References

Allardt E (1993) Having, loving, being: an alternative to the swedish model of welfare research. In Nussbaum M, Sen A (eds) The quality of life. Oxford University Press, New Delhi, pp 88–94. https://www.oxfordscholarship.com/view/10.1093/0198287976.001.0001/acprof-9780198287971-chapter-8. Accessed 10 May 2020

Banerjee R (2017) Does time matter? – a study of participation of women in urban governance. Space Cult India 4(3):62–73

Bittman M, Wajcman J (1999) The rush hour: the quality of leisure time and gender equity. SPRC discussion paper no. 97. Social Policy Research Centre, pp 12–17. https://www.researchgate.net/publication/45348041_The_Rush_Hour_The_Character_of_Leisure_Time_and_Gender_Equity Accessed 29 Dec 2014

Budlender D (2007) A critical review of selected time use surveys. Gender and development programme, programme paper no. 2, UNRISD. 7–33. https://pdfs.semanticscholar.org/6ded/5f661b9ba554c7a3cad113893d7cdcb7ab28.pdf. Accessed 29 Dec 2014

Department of Economics and Social affairs, SD (2005) Guide to producing statistics on time-use: measuring paid and unpaid work. United Nations, New York. https://unstats.un.org/unsd/publication/SeriesF/SeriesF_93E.pdf. Accessed 11 Sept 2017

Drage J (2001) Women in local government in Asia and the Pacific: a comparative analysis of thirteen countries. Summit of Women Mayor and Councillors: United Nations Economic and Social Commission for Asia and the Pacific. www.ucl.ac.uk/.../UNESCAP_Drage_Women_Local_Government_Asia_Pacific.pdf. Accessed 9 Mar 2018

Europen Communities (2004) Guidelines on harmonised european time use surveys. Working Papers and Studies. European Commission, Luxembourg. ISBN 92-894-8169-2. https://ec.europa.eu/eurostat/documents/3859598/5884753/KS-CC-04-007-EN.PDF/03057369-0bfe-47d5-b584-be0868d65f29. Accessed 10 June 2020

Ghosh A, Lama-Rewal ST (2005) Democratization in progress (women and local politics in urban India). Tulika Books, New Delhi

Ghosh J (2009) Transcending gender reservation, representation and identity among elected women representatives in municipal politics. Published Master of Science dissertation. School of Interdisciplinary Area Studies, University of Oxford. www.southasia.ox.ac.uk/.../Ghose%2C%20J%20MSc%20thesis%20June%202009.pdf. Accessed 1 Apr 2013

Hanson S, Pratt G (1995) Gender, work and space. Routledge, London

Harpreet D, Anupama U (2014) The time use survey of rural Punjab: measuring the unmeasured Rotterdam. IARIW 33rd general conference, Netherlands

Haug M (2012) Women's political participation at the sub-national level in Nepal: a mapping study (UN WOMEN). South Asia Country review series on women's political leadership. Norwegian Institute for Urban and Regional Research; UN Women, Norway, pp 5–20. https://asiapacific.unwomen.org/en/digital-library/publications/2012/6/womens-political-participation-at-the-subnational-level-in-nepal-a-mapping-study. Accessed 30 Apr 2013

Hirway I (1999) Time use studies: conceptual and methodological issues with reference to the Indian time use survey. International Seminar on Time-use Studies, Ahmedabad. www.undp.org/.../india/.../time_studies_conceptual_methodological_issues_reference. Accessed 17 Dec 2014

Hossain M (2012) Influence of social norms and values of rural Bangladesh on women's participation in Union Parishad. Indian J Gend Stud 19(3):394–412. https://doi.org/10.1177/097152151201900303

Keall M, Baker M, Montgomery H (2007) Time use survey scoping paper. Statistics. New Zealand, pp 1–25. http://archive.stats.govt.nz/~/media/Statistics/browse-categories/people-and-communities/time-use/tus-2009-10/tus-scoping-paper.pdf. Accessed 30 Dec 2014.

Lahiri-Dutt K, Sil P (2014) Women's 'double day' in middle-class homes in small town India. Contemp South Asia:2–15. https://doi.org/10.1080/09584935.2014.979762

Panday PK (2008) Representation without participation: quotas for women in Bangladesh. Int Polit Sci Rev 29(4):489–512

Pandey RN (2000) Sampling issues in time use survey- Indian experience. Working paper, Country paper for the expert group meeting on time use survey, New York

Rajivan AK (1999) Policy implications for gender equity: the India time use survey, 1998–1999. International seminar on time use surveys, Ahmedabad. https://millenniumindicators.un.org/unsd/demographic/sconcerns/tuse/Country/India/sourceind99a.pdf. Accessed 21 June 2016

Rizvi SH (2006) Empowering women through reservation in urban local bodies-difference between theory and practice. In: Singh UB (ed) Empowerment of women in urban administration (experiences and strategies). Serials Publications, New Delhi, pp 193–211

Task Force on Time Use Surveys (2013) Guidelines for harmonising time use surveys. United Nations Commission for Europe. Conference of European Statisticians, Luxembourg. https://www.unece.org/fileadmin/DAM/stats/documents/ece/ces/2013/TimeUseSurvey_Guidelines_for_consultation.pdf. Accessed 24 Dec 2014

The Community GPI Survey (Part II) (n.d.). http://www.gpiatlantic.org/pdf/communitygpi/communitypart2g.pdf. Accessed 31 Dec 2014

UNDP (1995) Human development report, 1995: gender and human development. Oxford University Press, Delhi

Chapter 6
Freedom and Functioning of the Women Councillors

According to the capability approach, 'functionings' and 'capabilities' are the best metric for most kinds of interpersonal evaluations. In other words, those interpersonal evaluations should be conceptualized in terms of people's functioning (their actual beings and doings) and their capabilities (the real opportunities they have to realise those functionings). These beings and doings together are held to constitute what makes a life valuable.

– The Capability Approach, Stanford Encyclopedia of Philosophy (2011)

6.1 Introduction

Amartya Sen's (2000) capability approach has already been explained in an earlier chapter. The capability has two kinds of variables: freedom and functioning. By using these two variables, one can judge the capability, that is, being and doing of a person. It is also a reflection of freedom to achieve valuable functioning. Freedom is considered an independent variable, and functioning is the dependent variable. In Sen's (2000) perspective, instrumental freedom has a consequential impact of bringing about other freedoms, or the development of society.

Instrumental freedom can be of five kinds. These are political freedom, economic opportunities, social opportunities, transparency, and protective security. However, the application of instrumental freedom to achieve valuable functioning is provided by the society, that is, substantial freedom. Achieving substantial freedom means the end of the development of society (Garrett 2005; Yupanqui 2011). The present chapter analyzes the instrumental freedom of the women councillors in urban governance, and how far it influences their capabilities to achieve valuable

R. Banerjee, G. Samanta, *Negotiating Terrain in Local Governance*, Local and Urban Governance, https://doi.org/10.1007/978-3-030-60663-3_6

functioning and to generate substantial freedom. The study of freedom of the women councillors in four cities of West Bengal also highlights the differences of the socio-cultural set-up of the cities, and its relation with the nature of freedom of women councillors.

6.2 Political Freedom

Political freedom determines the power enjoyed by the elected women in governance. The indicators to measure the level of political freedom are (a) access to control over the political space and government schemes, (b) position held in other organizations, and (c) support from members (Sen 2000; Joseph 2011).

Access is essential for women to make meaningful progress in each field. Therefore, equality of access to political space is essential for them. Such access to the political space can be measured by variables such as the nature of elected seats of the WCs (women councillors), that is, whether reserved for women or unreserved; tenure and work experience of the WCs in the governance field; position of the WCs in the municipal system, that is, whether they hold the position of Chairman-in-Council (CIC) or not; how much support they get from the political party from which they have been nominated; whether they belong to a family having a political background; nature of their political engagement; and decision-making abilities (Sen 2000; Joseph 2011). On the basis of these variables and indicators, the nature of political freedom of the WCs is analyzed both quantitatively and qualitatively in the four cities of West Bengal for in-depth understanding of the nature of freedom of the WCs.

6.2.1 Case of Darjiling

In the case of political freedom, it is found that among the interviewed WCs, 46 per cent are elected in unreserved seats (Table 6.1). This represents the people's faith in the capability of WCs. This faith has also been observed during the field survey in the city on many other occasions. Political parties have offered the tickets to women not only for the reserved seats, but also for unreserved seats.

Most of the WCs in this city have been elected for the first time with no experience in this field. It thus presents a good sign for bringing opportunities to the beginners. However, one of them had a background in politics during her student life at Siliguri College. She became a self-made administrator from a very young age. Two WCs are experienced in the work of councillor, as they have been elected for the second time. Of these, one is a CIC councillor of the water department (WC 29), and has been elected from the same ward for the second time. Another one (WC 14) was in the *Congress Sheva Dal* as chief executive, and was elected the last time too. She is an influential candidate from a different political background, that is, Indian

Table 6.1 Women councillors of Darjiling

WCs of ward number	Nature of elected seats	Nature of responsibilities
1	Unreserved	General
2	Unreserved	General
3	ST women	General
5	Unreserved	General
7	ST women	General
10	Reserved for women	CIC (land) now in waterworks
13	ST women	General
14	Unreserved	General
17	Unreserved	CIC (UPE CELL)
20	SC women	General
23	Unreserved	CIC (UPE CELL)
27	ST women	General
29	Unreserved	CIC (water)
30	Unreserved	CIC(water)
31	ST women	General

Source: Field Survey, 2013–2015

National Congress (INC). As the system of reservation is rotational, after completion of the second term, the reserved seat becomes an unreserved one. However, the WC has got the opportunity to work for the second tenure[1] with her improved capabilities, and learning from the first term.

The **influence of the political party** is the door-opening determinant in the political field. In such a context, Khattak (1996) has suggested that the political party should be legally required to give a percentage of party tickets to women. The political party also makes crucial decisions about whether women should be allowed to contest from general wards (Nair 2012: 479). Here, most of the WCs who were interviewed stated that they belong to the Gorkha Janmukti Morcha (GJM), and support the aim to have a Gorkhaland to secure the land for their future generations. The reason for their support is that the party has given them liberty to work for local people, as well as the transparency, guts, and dynamic behaviour of Bimal Gurung (the then chief of GJM party and Gorkha Territorial Administration (GTA)). One of them (WC 23) claims to view the party like her child, as she has developed it by her own hands.

WC 14 is from a different political background (INC), and said that she is not satisfied with her current political party; however, she has changed her party affiliation to the party in power in the city government. As the party is in rule, she has joined it. Although the councillor narrates that the party identity is less important to her than working for the common people, this observation indicates how the party controls the work of the councillor. She could not get much support from the urban

[1] <Footnote ID="Fn1"><Para ID="Par8">Second tenure signifies the second time of winning election of the candidate. One/first tenure is for 5 years and second tenure means a duration of 10 years as councillor from the same ward.</Para></Footnote>

local body as she was elected from another party, and that is why she changed her party affiliation to perform better in the neighbourhood. Similar findings have been cited in the work of Nair (2012), who stated that an independent candidate can change their party to sustain their political career, and to secure larger budgets and amenities for their respective ward. In Darjiling, all the women councillors are engaged in campaigning and meeting for the sake of their party, and for the larger demand of Gorkhaland. Some of them said that Mrs. Gurung (wife of the then Gorkha Territorial Administration Chief Bimal Gurung) encourages them to do their work.

While explaining the relation between the political engagement and the work of the councillor, all of them proudly confessed that the political party brings power, which is needed. WC 23 mentioned that her designation is first as party cadre, before being a councillor. The party gave her the ticket to be elected as a councillor. It is a positive sign that they all have a very clear vision about the reason for which they chose the political party.

Decision-making is an important component of political space for women. In this context, most of the WCs said that they take their own decisions. However, they discussed matters with the other councillors and in the ward committees. Except one, all of them said that they never discussed their work with their husband. The reason for this non-consultation is that it increases the burden on their husbands without giving any solution. They thus prefer to take their decisions independently without interference from their husbands. They do not feel shy while giving a proposal at a meeting. In case of placing decisions in the meeting, one general councillor[2] (WC 2) said that meetings are only meant for CIC councillors, as general councillors have no grounds to talk about the issues of their ward. Whenever she talks about her ward, other officials thump on the bench to interrupt and to stop her. Similar observations have been cited in other studies on similar questions as well. Ruet and Lama-Rewal (2005) observed in Kolkata that the disinterest of officials and the party influence determine the work of councillors. Palanthurai (2012) felt that such ignorance and ridicule is an obstacle for women's active participation. Another general councillor also confesses that she is often stopped midway while speaking by the other members in the meeting. Others are told that they should raise their issues before the meeting for the sake of the development of their ward. One CIC (WC 23) councillor said that she raises issues for all the wards in the municipality. She does not care about what others think about what she says in the meeting. Another CIC councillor (WC 17) from UPE cell said that she studies the agenda and gives positive or negative comments according to her own opinion. Thus, the finding is that though WCs are empowered here, they also face adverse situations in meetings. The reason is that they have their own agendas for discussion. For that reason, Holzner and Wit (2003) have stated that weak women have a better chance of becoming councillors than the stronger ones, who are not open to listening to the

[2] <Footnote ID=”Fn2”><Para ID=”Par13”>General councillors are those who are not members of the chairman in council (CIC), the cabinet of the local <IndexTerm ID=”ITerm115”>government<!--<Primary>Local governments</Primary>--></IndexTerm>.</Para></Footnote>

party leaders, who are men in most cases. Women who are weak are more easy to control by the party people, who are men in most cases, and they prefer to elect those women rather than the stronger women. However, the gap between general councillors and CIC councillor is very high in Darjiling, and general WCs always feel discriminated against because of their position.

Most of the WCs are quite aware of the sanctioned **government schemes** in the city. There was only one councillor (WC 13) who said that she has no knowledge about government schemes and their implementation. All of them are dissatisfied with the fund available for government schemes. The schemes are mainly sanctioned for the development of *jhora*, *nala*, footpaths, roads, walls, toilets, and slum area house development. The councillors act as a medium between the beneficiaries, contractors, and the municipal authorities.

Most of the WCs have **political membership** of the Nari Mukti Morcha. The president of this organization is Kabita Pradhan; and Mrs. Gurung also headed this organization. Eight councillors among them have no such membership to other political organizations. WC 2 has also mentioned that she has no time to do work for any other political organization.

Thus, in Darjiling, quantitative representation of WCs is very high, and six of them hold powerful positions in the Chairman-in-Council, and have responsibilities in different departments. WCs have very strong and clear thoughts about their political engagement. The WC from the opposition political party background (Indian National Congress) also changed her party to work for local people. In case of decision-making, they never discuss it with their husbands, and take their decisions on their own. There is a huge gap between the general and CIC women councillors in terms of power. CIC councillors control everything, and general councillors have no power. Though the quantitative representation is high, there is conflict in order to retain power, and to be more powerful by suppressing others. As a result, grouping and multipolarity of power are found between them. Most of the WCs are not satisfied with the available funds for the government schemes, and one of them has no knowledge about government schemes. All these factors hinder their substantial freedom. We did not find any proxy councillor in this city.

6.2.2 Case of Balurghat

In Balurghat, all the WCs are elected under reserved seats for women, except the WC who is at the post of chairman (39%). There is only one woman, that is, the chairperson (WC 1) in the Chairman-in-Council position (Table 6.2). However, all the WCs are very active in doing their work. The experience of the WCs about their political space is contradictory. For example, one WC has new experience as she has been elected for the first time (this is the case for four among them, of which two are below 30 years, and two are widows). Other WCs (three, including the chairman)

Table 6.2 Women councillors of Balurghat

WCs of ward no.	Nature of elected seats	Nature of responsibilities
1	Unreserved	CIC
7	Woman	General
10	Woman	General
12	Woman	General
16	Woman	General
17	SC woman	General
18	Woman	General
21	Woman	General

Source: Field Survey, 2013–2015

have the experience of having been a councillor two to six times already.[3] Thus, there is an amalgamation of experienced as well as beginners here. It is a good sign, as the beginners get the opportunity to learn about their work from the experienced ones. In this context, literature says that women are more effective in the long run than men in governance (Sathe et al. 2013: 54). This is also proven in case of the effectiveness of the women councillors of Balurghat. For example, WC from ward 21 (WC 21) has developed her space in her ward, as her husband was elected two times as a councillor from the same ward, and she was elected the last time against reserved seat for women. This case represents experienced, but proxy representation of the WC to protect the seats of their husbands. Similar observation is also found in the study of Ghosh (2003) and Singh (2006). They observed that women are elected to reserved seats at the mercy of their relatives (husbands, father and father-in-laws) for securing the seat for their next relative, as their relatives have association with political parties.

All the councillors belong to the Revolutionary Socialist Party (RSP).[4] The reason behind their commonality is the party principles and political family background. In that context, WC from ward 18 mentioned, ‘I am called as Students’ Federation of India (SFI) councillor as I was doing SFI in college’. **Party support** is an important determinant which helps women in making their way into the political world. However, the perception of all the councillors regarding the support from their party is not the same. A younger member (WC 18) says that the party does not help her. Then, she explained this, saying that politically the party relied on her and that is why they gave her liberty. On the other hand, the chairperson, as an age-old supporter of the party, opined that the party helped bring her to her position, and though there may be a difference in opinion it must be established with logic and then accepted. As stated in the literature, women have to make a place for themselves in the political party to share power (Gupta, 2006). From this point of

[3] <Footnote ID=”Fn3”><Para ID=”Par18”>Two times mean for two <IndexTerm ID=”ITerm146”>tenures<!--<Primary>Tenures</Primary>--></IndexTerm>, i.e. for 10 years; and 6 times signify for 30 years.</Para></Footnote>

[4] <Footnote ID=”Fn4”><Para ID=”Par20”>RSP<IndexTerm ID=”ITerm156”><!--<Primary>Revolutionary Socialist Party (RSP)</Primary>--></IndexTerm> represents one partner of the Left Front in West Bengal.</Para></Footnote>

argument, it is clear that the party nominated the women in the reserved seats not only for proxy, but also to work for the citizens independently. Panday (2008) suggested that it is a good measure that political parties should take some initiatives for gender equality. Most of the WCs have their individual reasons for their support of the political party.

Decision-making is the quality that determines the status of self-dependency of the participants in a system. To increase such decision-making abilities among women, Sheel (2003) has suggested a law regarding a gender-equal society in the context of Japan. In Balurghat, all the WCs are more or less independent in decision-making, except for the unmarried one. The WC from ward 9 said that she discusses matters only with her young daughter who also helps her in giving certificates to the people during her absence. Another WC from ward 21 discussed matters with her husband as he was an experienced leader. In other words, her husband though not in power used to administer the system. This fact becomes unconsciously visible in her words. The young WC from ward 18 also discusses matters with her father. However, all the councillors discuss their decisions with the members of the ward committee as well as the board of councillors. On the question of having to take decisions, all the WCs agreed that there is sufficient space for putting forward their decisions in the meeting. That is why the problems have not been raised. Most of them acknowledged the influence of the woman chairperson for empowering and bringing them confidence. In this context, literature says that women have proved themselves as better team leaders than most men around them. They have a deeper sense of loyalty and commitment (Narain and Mridubhashini 2006: 186–187). Echoing almost similar words, the young WC from ward 18 mentions, 'I do not listen to anybody when it comes to making decisions and my ruling personality, egos and psychological confidence helps to force others to agree with me'.

The chairman, as a woman, may be said to obstruct other women councillors from holding the Chairman-in-Council (CIC) position. However, she encourages all of them to be empowered, and make individual decisions in the political field. In this context, as a woman chairperson, she has mentioned, 'My experience and tricks helped me to overcome the hurdles of opposition in case of giving any decisions'.

The awareness of the WCs regarding the **government schemes** is quite high. They take part in implementing, maintaining, and giving information and encouragement regarding the government schemes to the common people. However, young WC 3 does not know the details as the work is fully administered by her ward development secretary. This shows the proxy representation of the women representatives in the governance field. She said that her ward secretary has better information about her ward. In the case of fund availability, everyone said that more is always welcome.

Other kinds of **membership** in the political field help in spreading the contacts of representatives in the political sphere. However, most WCs have no such membership with any other political organization, though young WC from ward 18 has membership of the youth association, and the chairman is a member of the RSP central committee.

In Balurghat WCs are elected against the reserved seats only. Although the chairman is a woman, there is no other woman in CIC position. Some women members are experienced, and they are elected for two terms, whereas some others are new to the post. Only one woman holds the position of men's proxy. Most of them were attached to a political party from their student life and have prior affinity to it due to their family background, or due to engagement in political life during college days. The political party has nominated those women for councillor positions who are either widowed or unmarried. The purpose is to enable them to work properly without the extra burden of their family. The woman chairperson shows considerable solidarity to help women enjoy their power and substantial freedom, besides encouraging them in taking decision in the meetings.

6.2.3 Case of Raniganj

The representation of WCs in terms of total councillors is 45.45% in Raniganj. Among the seven interviewed WCs, one holds a position in the Chairman-in-Council (CIC), while the others are members of the board of councillors (Table 6.3). Among them, five are elected to the reserved seats for women, and only two are elected to unreserved seats. Ghosh and Lama-Rewal (2005) in their study in Kolkata stated that those who are elected to reserved seats are very unsure in saying whether they will be elected if the seats become unreserved the next term. It hinders their enthusiasm for good work. From the field observation, two viewpoints are very important and must be mentioned here. WC 12 has said, 'When the seat becomes reserved, everyone approached me and I came to the seat to see how it is'. In this context, the ward secretary of the WC 1 added that there was a qualified man for the post if it had not been reserved for women. Nair (2012) argued that political parties have to select women for the post of the councillor to obey the constitutional amendment. Because of protocol, women are permitted to stand in the election and attend the meetings. It signifies that reservation may assure the numerical representation but not the active participation of women. It depends on other factors also (Bharathy

Table 6.3 Women councillors of Raniganj

WCs of ward no.	Nature of elected seats	Nature of responsibilities
1	Women	General
4	ST	General
9	Unreserved	General
12	SC woman	General
13	Unreserved	CIC
14	Woman	General
17	Woman	General
21	Woman	General

Source: Field Survey, 2013–2015

2012: 81–82). In Raniganj, party members have no wish to elect a woman to the seat of the local representative. WCs who are elected to reserved seats are also elected for the first time. In this context, Thakkar (2006) has mentioned that political parties do not consider women as a valuable constituency, until women can be made into a vote bank.

Ghosh and Lama-Rewal (2005) have stated that the majority of women are newcomers without political education, as the quota system only appears to have been thrust upon them without giving enough time and opportunities to prepare for this considerable responsibility. Only two of them, who are elected to unreserved seats, have prior experience of contesting in the elections. WC 13 has been elected twice from the Communist Party of India (Marxist). Rizvi (2006) stated that lack of experience (first or second time) of elected women hinders their active participation in decision-making. WC 9 who is from the opposition party, All India Trinamool Congress (AITC), was elected after contesting with four competitors. She had also contested for the seat of Member of the Legislative Assembly (MLA). She has a long history of contesting in elections. In her words:

> I am a housewife of a backward family (Muslim). I contested against the famous leaders of the opposition party and tolerated many tortures. For example, I was in jail in 2009 for eleven days and listened to many negative comments about me from the members of the opposition. In 2010, I won the election by securing the demands of Below Poverty Level candidates and common people. I learned Hindi to better communicate with the people of my locality as most of them are Hindi speaking people.

From her words, it is clear that she has achieved her position after long-term political struggle, and she has not been selected but elected by the people of her ward due to her contributions. In the context of Bangladesh, Panday (2008) identified that one reason for less political freedom of the women representatives is the strong Islamic ideology, which reduces the chances of women to participate in the political space. However, this is not a hindrance in the way of the Muslim woman councillor in Raniganj. She even learned Hindi to facilitate better communication with the people of her ward. It represents her eagerness to work for the people and develop her identity as a local representative.

In case of political party-wise representation, WC from ward 9 is from the All India Trinamool Congress. WC 21 is from the Communist Party of India (CPI), and the rest are from Communist Party of India (Marxist). All of them acknowledge the support they get from their parties. WC 21 who is from CPI party worked as a woman secretary of the party, before becoming a councillor. She thus had experience in the political field before becoming a councillor. However, recently her work was being obstructed by her party members because the chairman of the municipality gave more priority to the party members than her. He (chairman) discussed with the party members of her ward about any developmental work, even though she remains present in the municipal office. As a result, she has faced a hidden power struggle, and identity crisis. Such opposition from their own party members disturbs the women councillors, especially in the metropolitan cities. According to Ghosh and Lama-Rewal (2005), this kind of opposition may be for two reasons: (a) men feel threatened when women come up; (b) men are resentful as they have to give up 'their' seats for women because of reservation.

WC 12 has a different viewpoint regarding the political party. She told her party members that she would work either for the party or do the work of the councillor. After that, she has only been doing the work of a councillor irrespective of the identity of any political party. She also added that the opposition party does not create any problems in her work as she works for them too. Her only objective is to work for everyone in the ward. The findings here are thus very different, where a political party nominated a woman candidate and she does the work of the councillor, not for the party's sake, and that too without any influence of the political party. In Pakistan, Shah (2015) also stated that women members of the national assembly (MNAs) are active participants though elected through quotas as they want to prove their capabilities to the political parties. Here, it represents the openness of WC 12's party members as well as the determination of the WC about her decision. It is a prerequisite for good governance. In Kolkata, Ghosh and Lama-Rewal (2005) said in this context that women of the Communist Party of India (Marxist) cadre are assessed by their popularity, clear image in society and some political principles, but are not identified by their political party affiliation.

In the case of **party support**, WC 9 from All India Trinamool Congress party said that her party can do anything to remove the obstacles in doing work. However, in the case of the control of the party, she said, 'I am always trying to neglect controversies. But the individuals have a different mind that is why it is not possible to be good to everyone. I am a woman from minority and struggling for the people. I feel proud of that'. This represents her good leadership quality. Even after considering all the difficulties from party members, she remained firm to work for the people.

Four councillors belong to families with a political background, while the others are beginners. One Chairman-in-Council and two general WCs stated that their father-in-laws' family had a political background, while the father of WC 1 was in politics. In the case of membership in other political organizations, only three of them have membership of the *Mahila Samiti*, while the Chairman-in-Council is also a member of Communist Party of India (Marxist).

Decision-making means the freedom to express an individual's own viewpoint. All the councillors have said that they put forward proposals in the board of councillors meeting, and sometimes express their opinions in a verbal form. There is no problem of killing agency and the environment of the municipality is very supportive. However, three viewpoints are very important to mention here.

WC 12 who only works for governance, said:

> If the necessary work written in the proposal is not sanctioned for my ward within time, I become vocal in the meeting as I am elected to do the work for the people.

This shows that she is concerned about her ward development and about her work as she has been elected to do so. This ethical stand about accomplishing their own work is essential for everyone, and clearly is very high in her.

CPI WC 21 in this context has given a long argument and releases her frustration towards the municipality. She said:

> As a woman ward councillor, I do not get any facilities and do not get the work as I want. That is why I always feel sorry. I gave my written proposal to the Chairman; however, the written work of the ward secretary is approved by the Chairman. I am not important for them.

As she is from the opposition party, and she is a woman, it is easy for the chairman to make her invisible. Such kind of negative attitude on the part of municipal officials is one of the major obstacles for women leaders in governance (Palanthurai et al. 2009: 63). The chairman gives more priority to the ward development secretary and other party members. She thus has no voice and remains a namesake councillor who holds a reserved seat. Therefore, just the presence of women in governance is not a guarantee that they will be valued and that their interests will be protected. However, a more proportionate number will at least legitimize their voice collectively (Beall 1996).

In continuation, she has also said:

> Other WCs have got the house from the Municipality's fund given by Chairman. However, I do not get a Municipality's paid water connection. I placed my problems in front of Chairman many times. The Chairman remained accommodative and told me from last two years that he would try. For that reason, I remained silent after seeing illegal activities and I have not attended Board of Councillors meeting for the last two months. I also argued that I will resign if the work of my given proposal is not implemented. However, still the work is on the paper.

From these arguments, it is clear that she has staged a silent protest many times to assert her identity and position. However, the chairman does not listen to her. As a result, her identity is in crisis though she is in position and wants to work for the ward people. The study of women leaders in Australia by Sheridan et al. (2011) noted the subordinate position of the women leaders and stated that they remain invisible and suffer from this space of betweenness, that is, between their formal space and the invisible space. However, in the present study, WC from ward 9 who is from opposition party (All India Trinamool Congress) has said in a dominating manner:

> I do work for a total of 22 wards. I just take a look at the proposal and the party members inform me if there is a work in the Municipality.

On the matter of killing agency in the workspace, she (WC 9) has said:

> My seat is in the last corner at the Board of Councillors meeting but I do struggle to maintain my identity and form a group with the departmental members of the Municipality and have fought for the issues such as development of bank account for salary, increase of the salary and development of provident fund facilities for them. They have come to join this group by their own wish, not by pressure.

This phenomenal action of the woman represents her good decision-making abilities as well as influential leadership qualities. Despite being in the opposition party, she has fought vehemently to make her own identity and for her own betterment as well as to fulfil the demands of the other members of the municipality. This action might be considered from the standpoint of political structure of the state and the control it has on the decision-making of the local body. The woman probably shows more power as her party is in control of the state government under which the municipal government works even when they are run by the opposition party like the Communist Party of India (Marxist).

All the WCs are partly satisfied and sometimes unsatisfied with the work of the **government schemes** and fund availability. As councillors, they give a proposal to the chairman, and then it would be accepted on the basis of priority. After that, during the phase of implementation, they act like the manager. WC 1 proudly elaborates about her work and says:

> I have completed the work related to the development of Amodpur drain and saved the locality from severe flood. I also developed a bridge in Mongalpur. I have also opened bank accounts for labour class people of my locality under the Government scheme of 'Bhaviswatnidhi Prakalpa'. However, I am quite disappointed for non-completion of the work of a road of my ward.

This fact revealed her honesty and strong wish to work for the people as well as for development of the ward.

However, the CPI councillor (WC 21) argued the hidden facts of her identity crisis regarding the implementation of government schemes. In her words:

> As a woman ward councillor, I do not get any facilities and do not get the opportunities to do work as I want. That is why I always feel disappointed. I gave my written proposal to the Chairman; however, the work given by ward secretary is approved by the Chairman. I am not important to them. I had given the name of the beneficiaries who has urgent need to get houses after doing extensive field work of my ward. However, the ward committee has changed the name and they get benefited. Ultimately councillor is just a position which has no value.

WC 17 in a protesting manner added: 'No work can be called complete; as I developed one side, on that time another side becomes deteriorated'.

WC 12 responded in the voice of common people that:

> I have told the boys who are engaged in cleaning that you are also a poor human like me; please do your work properly. I don't like to say hard words to you and also do not make complaint against you.

She (WC 12) always represents herself such as a leader of the common people. This is a very good characteristic for doing better in the governance field. The distance between local representatives and common people will increase if there is a feeling of vanity among local representatives due to their position. It could be very harmful to governance.

However, All India Trinamool Congress (AITC) WC 9 complained regarding the implementation of government schemes by the municipality. She said that the money for the schemes goes back to the treasury because of non-implementation. She developed 14 self-help groups and gave 14 lakh (1.4 million) rupees in total to them.

With all the controversies, it is clear that most of the WCs have the courage to work for the development of their wards. However, the CPI and AITC WCs are trying to implement these schemes in their own way. They protest against the internal rules of patriarchy in the municipality.

WCs have no **membership** in other political organizations because of the inherent patriarchy in the municipality.

In Raniganj, there is only one WC in Chairman-in-Council (CIC) position who is fully controlled by the chairman. The WC who is in CIC position tries to control other WCs. However, two of them try to sustain their own identity as councillors. The orthodox nature of political views regarding women in the political party results in nomination of only those women who belong to reserved categories and those women who have no prior engagement with the party before becoming a councillor. The WCs from the opposition party face severe discrimination in the municipality. They try to protest against it to retain their identity as councillors.

In case of decision-making, only the voice of WC who is in CIC position has been considered. However, other WCs are also struggling to make their demands heard for their own wards. The WCs of the opposition party stopped attending meetings because of discriminatory practices. All of them take part in the implementation of government schemes. However, most of the government schemes are implemented by the ward secretary or by the party members. WCs in that case have no work to do, and if they wish to do something, they find many obstacles in their way. Political freedom is restricted here, and the WCs are treated as though they are useful only for acquiring reserved seats. Although women are elected in proxy seats, they never remain silent and struggle in their own ways.

6.2.4 Case of Hugli-Chinsurah

Among the WCs, two have experience of being elected for two tenures. There is only one WC in the CIC position (Table 6.4). She has the responsibility of the finance department. However, another experienced WC of ward 16 has protested

Table 6.4 Women councillors of Hugli-Chinsurah

WCs of ward no.	Nature of elected seats	Nature of responsibilities
1	SC woman	General
2	Woman	General
5	Woman	General
9	Woman	General
13	Woman	General
16	Woman	General
17	SC woman	General
20	Woman	General
23	Woman	General
25	Unreserved	CIC
27	Woman	General

Source Field Survey, 2013–2015

against the system of the municipality, where there is only one woman in CIC position. She has argued:

> I have the experience as a councillor for two terms. However, I have no chance to get CIC position which the inexperienced men councillors have. I have told the Chairman many times but he simply ignored.

Except for one, all other WCs have been elected to the seats reserved for women. The nominated WCs of Hugli-Chinsurah are found in four types of groups. One group of WCs was elected because of the popularity of their husbands/men members of the family in the ward, or if the men are active members of the party. The second group of WCs are securing the seat for the men members of the family. In most cases, the member is the husband, and in other cases, he is the brother-in-law. In the third group, WCs are fully guided by their male party members, and in the fourth group, WCs are struggling to retain their identity through their own work.

In the first group, there are three WCs. Their husbands or brothers-in-laws are very popular in the ward as well as active party members. This category of leadership is considered a dangerous trend in governance as stated by Singh (2006) and Joseph (2011). They also help the women by doing her work as the councillor. For example, they undertake ward visits, and even write letters using the councillor's pad on which the WC just puts her signature. In this context, the popularity of the husband acts as a positive factor for the women to be elected to the position of councillor. However, the continuous helping nature of their husbands for doing work of the councillor reduces the burden on the wives, making the seat as 'partly proxy' in nature. This pushes us to rethink the concept of proxy women. According to Nair (2012), women who flaunt their power, but also lean on men for legitimate support are like proxy women. In this context, Palanthurai (2012) stated that there is also an official tendency to include the men of the elected women representatives' family to do the work of the local representatives, and cultivate corrupt activities through the men of their family. This also results in the elected women becoming proxy.

In the second group, there are again three WCs. Among them, WC 9 is securing the seat for her brother-in-law who was the councillor as well as chairman of the municipality for 25 years. WC 20 was elected from opposition party, i.e. Forward Block, and is securing the seat for her husband for the next term. Her husband was also the councillor in the previous term from that ward. WC 23 is also securing the seat for her husband. He was the councillor for the previous two tenures of that ward. This group of WCs is elected to proxy seats of the male members of their family to secure the seat of the councillor for the men of the next generation. Panday (2008) has also highlighted that proxy nomination of women representatives by political parties is also a major barrier for women's political participation in decision-making in Bangladesh. Thakkar (2006) in his study of Kolkata stated that women councillors are considered as 'men's political resources'. In Kerala, women are used as tokens in *Pani Samitis*; sometimes they do not even know why they are elected as observed by Kulkarni (2011). From his study, Palanthurai (2012) also identified that political rulers treat women candidates as a key to control the governance system. However, John (2012) has a different observation, and stated that though during this time WCs have received some knowledge about the work of the councillor and are able to do some work on their own, still they know that the party will never nominate them when the seat becomes unreserved in the next tenure. Then, male members of their family will be nominated for the same seat. This fact also brings an inherent hopelessness among them about their work. Such rotation of political career not only affects the beneficiaries of the quota, but also diminishes their accountability towards voters. This phenomenon is more pronounced in the cities where political parties play a bigger role in local elections, as observed by Ghosh and Lama-Rewal (2005).

In the third group, WC 2 is fully guided by the male members of her party. She acts like a proxy woman as she does not have the opportunity to make her own decisions in case of development of her ward. Singh (2006) and Joseph (2011) have identified this category of women's leadership as harmful when party members use the power of the respective elected women. In the fourth group, WC 16 has two terms' experience as a councillor, which she achieved by her own work. However, she is discriminated from a seat in Chairman-in-Council. WC 17 has been elected to the seat reserved for 'scheduled caste women' for the first time, but she has done developmental work and visits the ward by herself without the help of her husband. The choice of a political party to nominate a women candidate is an influential factor for women's entry into governance, as stated by Panday (2008). There are very few educated women who get elected and contribute to the well-being of their locality. However, they are sensitive and positive about their work (Singh 2006).

In case of **political party** affiliation, four of them are from the Indian National Congress (INC) background. At present, they have shifted allegiance to the All India Trinamool Congress (AITC) party. Among them, the grandfathers of WC 16 and 23 were freedom fighters during India's freedom movement. The reason for their shifting parties is the principles of the main political leader of the AITC party, Mamata Banerjee. They attend the meetings and rallies with banners for the sake of the party. All of them have acknowledged the support they get from their party. In this context, WC 2 said:

> The party's male members do all of my work and I hold just the position. During any occasion of the party, all the members come to my house and prepare banners and do other things. They also remain all the day in the office at my house to do different works of the councillor.

From this argument, it is very clear that the WC has very little knowledge about her work, as all the work is done by the party members. She just remains to hold the seat as it is reserved for a woman. Otherwise, she would never have been elected to the seat. After the election, she has not even got the opportunity to know more about her work, and she never feels disappointed about that, as she feels that party members release her from the huge burden of her work. WC 23 argued:

> Party members supported me during my husband was in the hospital due to serious health problems. However, I have told them many times to give the job to my sons. They did not listen to that. They support me in many times but never gave job opportunities to my sons.

From her confession, it can be seen that she has been elected to the proxy seat not for securing the seat for her husband but for getting the jobs for her sons. She is elected to fulfil the wishes of the male members of her family, both present as well as future generations.

WC 20 is the only member of the opposition party, Forward Block. She is inspired by her father, who was a member of Communist Party of India (Marxist); and she was a Students' Federation of India's active student member during her college life. She has acknowledged the support she gets from her party. She also added that the members of the All India Trinamool Congress party asked her to join their party. However, she ignored them as she felt it was ethical not to change her party, even if she failed to win the election.

Decision-making is an important determinant of freedom. However, most of the WCs are elected to a proxy seat of their husband or brother-in-law. In most cases these male members do all the work of the councillor, and she just puts her signature on it. The proxy in the election process, meetings and decision-making has caused serious damage to women's empowerment. Like their husbands, the in-laws of the women councillors are the main decision-makers, relegating real women representatives to the background as observed by Singh (2006) in his study. In the meetings, they have to present themselves to talk about the problems of their ward. The board of councillors meetings occur once a month, and a car is provided to all the WCs to reach the municipality on meeting days. Paradza et al. (2010) suggested that municipal transport is an important recommendation for the betterment of work. However, they have been told that as the car follows a detouring path to cover all the WCs houses, it's better to go at their own cost. This will save time and they can leave their house at their own timings. In the context of the meeting, all of them have said that they verbally speak about the problems of their ward and, in some cases, they have to write a proposal on their councillors' pad and, on the basis of priority, it is sanctioned. In some cases there is a resolution about work and they discuss it in the meetings. In this context WC 9 has said that she has been oppressed sometimes as there are many members in the meeting. The WC from opposition party (ward 20) has said:

> They never create any opposed situation when I talk at the meeting but they never listen to my words. I know whether I say or not is the same thing.

There is discrimination in the meetings on the basis of affiliation to political parties, as observed by Hossain (2012) and Khan (2014) in their studies of Bangladesh. In Bangladesh, the performance of women councillors was reduced to the role of listeners and observers at the Union Parishad meetings, as the male members were not even listening attentively; instead they would interrupt and laugh at them or even display anger. They usually would not accept the proposals given by the women representatives. As a result, women representatives could not attend the Union Parishad meetings. Palanthurai (2012) also stated that discrimination in meetings is an important obstacle as mentioned in the literature, and this is also proven here, though the discrimination in this case takes place mostly on the basis of political party.

The husbands and brothers-in-law of most of the WCs are in the political field. As a result, most of the decisions are taken by them (male members). In one word, WCs are guided by the men in case of decision-making. However, WC 17 discussed with other councillors and took her decisions independently. Her husband is a municipality surveyor, and has no time to discuss with her. She does everything on her own.

For the implementation of **government schemes**, WC 17 said that she has implemented many schemes such as *Gitanjali*, *Rajiv Awas Yojana*, etc. under Kolkata Metropolitan Development Authority (KMDA). When the schemes come up for implementation, she goes to every house in her ward to give them the relevant information. She also prepares a list of schemes in her own handwriting as she is unable to afford the cost of computer typing and printing. Then she makes photocopies of the original and makes posters for people everywhere. WC 25 said that as it is a non-slum ward, it was never facilitated by many government schemes, but has got some KMDA projects. WC 20 said that as she is from the opposition party, no schemes have been sanctioned for her ward. Other councillors implemented all the government schemes sanctioned for their ward. However, the work of communication with the common people regarding government schemes is done by the male members of their family, or by the party members.

In the case of **membership**, they have no membership to any other political organization. Their exposure to the political field is very low.

6.2.5 Overall Findings

From the values of composite scores, it is found that Balurghat reveals higher status than Darjiling. In Darjiling, quantitative representation of WCs is very high. The percentage representation of WCs elected to unreserved seats and holding the position of CIC is also very high in Darjiling (Tables 6.5A and 6.5B). However, most of them are inexperienced. Above all, though most of them have decision-making abilities, the lack of networking, and the gap between CIC and general councillors make their position disadvantageous when compared to Balurghat. On the other hand, the integrity among the WCs is very high because of the influence of the woman chairperson in Balurghat. Except for one, all of them are elected to reserved seats, but they are experienced in politics as well as governance.

In the two cities of south Bengal, the emancipation level of the WCs is low and runs in negative value. The most disadvantageous position for WCs is in Hugli-Chinsurah, as most of them are elected to proxy seats, and also remained proxy in case of decision-making. In Raniganj, the lower status of emancipation of WCs has come about because of the controlling nature of the chairman and discrimination on the basis of political parties. In this context, Horowitz (2009) has suggested the necessary requirements for women to get good governance—women-friendly institutions and gender-sensitive mechanisms.

Table 6.5A Political freedom of the women councillors

Indicators of political freedom	Darjiling	Balurghat	Raniganj	Hugli-Chinsurah	Mean	Standard deviation
% of women councillors elected in general seats (A)	35.71	14.29	25	11.11	21.52	11.16
% of experienced women councillors (B)	14.29	42.86	12.5	22.22	22.96	13.91
% of women councillors elected in Chairman-in-Council position (C)	35.71	14.29	12.5	11.11	18.40	11.61
% of women councillors from influential member of political party (D)	14.29	42.86	25	0	20.53	18.06
% of women councillors from political family background (E)	14.29	42.86	37.5	22.22	29.21	13.24
% of women councillors elected in no-proxy seat (F)	100	85	100	22.22	76.80	37.07
% of women councillors who are in student politics (G)	28.57	57.14	0	11.11	24.20	24.90
% of women councillors who have decision-making abilities (H)	42.86	42.86	25	44.44	38.79	9.22
% of women councillors who actively participate in the implementation of government schemes (I)	42.86	57.14	50	44.44	48.61	6.45
% of women councillors who have the membership in other political organization (J)	14.29	14.29	37.5	0	16.52	15.52

Source: Field Survey, 2013–2015

6.2.6 *Gender Matters*

In spite of having one-third or more representation of women in the city government, the higher positions in the power hierarchy are still far from the reach of women. In four cities we have only come across one WC in Chairman-in-Council (the cabinet of the municipal government) position in Darjiling. Although there are others who are qualified, they are not given the position in Chairman-in-Council. The women are often either occupying proxy seats to hold their husband's seat as long as it comes under women's quota, or they got elected because of the popularity of the men of their family. Their identity remains politically overshadowed because of these factors.

Because of the social norms and internalization of patriarchy, women also welcome the help by their family members, which in a way reduces their additional burden. A proportion of them still feel that it is their own job, and they can do it on their own and feel reluctant to take the help from other family members. Like Omvedt (2005) stated, the idea that women's place is in the home still defines

Table 6.5B Political freedom of the women councillors

Cities	Z scores of (A)	Z scores of (B)	Z scores of (C)	Z scores of (D)	Z scores of (E)	Z scores of (F)	Z scores of (G)	Z scores of (H)	Z scores of (I)	Z scores of (J)	Composite score of political freedom
Darjiling	1.270	−0.623	1.490	−0.346	−1.127	0.626	0.175	0.441	−0.890	−0.144	0.872
Raniganj	0.311	−0.752	−0.508	0.247	0.625	0.626	−0.972	−1.495	0.215	1.351	−0.352
Balurghat	−0.648	1.429	−0.354	1.236	1.030	0.221	1.322	0.441	1.321	−0.144	5.854
Hugli-Chinsurah	−0.933	−0.054	−0.628	−1.137	−0.528	−1.472	−0.526	0.613	−0.646	−1.064	−6.375

Source: Field Survey, 2013–2015

women's position, and their role in the public space and work outside home is considered as secondary. However, some women are trying hard to create their own position within the municipal government and adjusting against all odds to retain their power and position. It has also been observed that given even limited support from home and office, they can perform much better. If the woman is from the opposition party, they often face marginalization on double counts—one as a woman and the other as representative of the opposition party (Samanta and Banerjee 2013).

6.3 Socio-economic Freedom

Socio-economic freedom is a combination of both social and economic opportunities that the elected women have access to. Social opportunities refer to the arrangements that society makes for education, healthcare, and so on. It influences an individual to live better. It is important not only for private life but also for effective participation in economic and political activities. It has a direct effect on a person's effective freedoms, such as social improvements and social interactions (Sen 2000; Joseph 2011). Economic facilities include the financial freedom of elected women representatives, mainly the right to property enjoyed by the men and women. Therefore, the socio-economic freedom includes education, physical mobility, training, confidence to speak in public, the vision of the future, family support, respect from the community and occupational status (Sen 2000; Joseph 2011). Palanthurai et al. (2009) opined that the main problems of women representatives in India occur due to sociocultural mooring in the system, which creates hesitation within women while participating in the men's space of administration. Sekhson (2006) stated that effective participation of women was limited by the barriers rooted in patriarchal social structure and norms about a woman's proper place and appropriate behaviour in society.

6.3.1 Case of Darjiling

The social background of the WCs who have been interviewed is described here. Most of them have a primary level of **education**. Two are graduates, four have studied till high school, and two continued their education up to class 8, while the remaining six have a primary level of education. In Pakistan, illiteracy of women acts as a negative factor for political participation. However, Khattak (1996) in his study of India stated that in India party members prefer to nominate women with low literacy, as it makes it easier to exercise control over them. Among the WCs, four were employed before becoming councillors, and they have continued with their jobs. Among them, two are teachers, while two are in private jobs. Others do business such as keeping paying guests, running a meat shop, etc.

Mobility to public spaces is an important empowering condition for women, which allows them to engage in gainful work and gives them access to other benefits available for them in the world outside their homes (Samanta and Sil 2013). Panday (2008) noted that women are less **mobile**, despite being in governance. However, WC 14 has visited places such as Delhi and Hyderabad for party reasons, and WC 29 has also visited Delhi. WC 23 went to Kolkata three times. The fact is thus different from the theoretical notion that women representatives are less mobile. Here in Darjiling, most of the WCs are mobile beyond their municipality and have visited many other places for participation in politics. It represents their independent mentality and breaks the stereotypical norms.

On the question of **training** and workshops, the surveyed councillors responded that the municipality does not offer any opportunity for training to the councillor. The other responses are very important to mention here. WC 14 has responded that there was a discussion related to training, but it was not implemented. WC 29 has learnt about her responsibilities from the departmental members. WC 23 has taken an initiative regarding the training from State Urban Development Agency. However, it was not implemented. In this context, Gupta (2006) suggested the importance of training, which should focus on the critical areas of urban governance such as environment, health, education and sanitation.

WC 13 mentions that she wants proper training to learn about her duties and to gain the confidence to speak in public. WC 5 responds that she does not need any training for work, as her work consists only of putting her signatures on paper. The institutional difficulties of the higher authorities obstruct the path of their empowerment at work by not providing any training to them. Palanthurai (2012) identified this lack of training facilities as an important obstacle for the women representatives in India, and this has been observed in Darjiling as well.

The confidence to speak in public partially depends on the family background as well as nature of engagement in politics. Here it is important to mention that only two of the WCs belong to a family having political background and have been engaged in politics actively from a very young age. The comparison between case studies will make the facts more visible. For example, as previously discussed, WC 13 who wants training to get the confidence to speak in public does not have a political family background, and she was not involved in student politics either. On the other hand, the father of WC 14 was a congress leader, and WC 29 has been engaged in politics from her childhood. She delivered her first political lecture at the age of 16 years on the platform of *Suguna Janasava*, and has been complimented as a good orator.

Everyone answered gladly and positively regarding the **support of their families**. They all say in one sentence that, 'without the support of their family, it would be very hard to achieve as well as maintain this position'. Most of them are married, and they have mentioned the name of their husband as a supporter. However, this is not the case for everyone. Four different observations can be cited. WC 13 said that her family is quite offended as she is working. WC 29 has stated that availability of ready and packet foods have helped her a lot in fulfilling her domestic duties like cooking. WC 31 mentioned the name of her *bhabhi* as a helper in daily chores.

However, there is an inherent patriarchy evident in the support of their husbands, which comes to the surface when WC 14's remarks are considered. She proudly says that her husband gave full support to her work, but imposes the norms to be followed as women, as he says 'wear a proper dress, behave properly and do your work like a lady'. The nature of family support also represents the patriarchal norms set for women. Nanda (2012) in his study of *Panchayat* of Rajasthan observed that women's dresses are also determined by the men of their families. This kind of domination of men in the family is a significant obstacle for women leaders as identified by Palanthurai et al. (2009) as well.

On the issue of **satisfaction, gaining respect** and **vision for the future**, everyone responded positively. WC 29 answered in a politically correct way that it depends on local people's expectations. One of the general councillors also said that though she is satisfied with her position, it is her first as well as last time here because of her family responsibilities.

WC 31 said that she does not like politics. That is why she does not want to pursue a political career. We received the same opinion even from women members who are engaged in Chairman-in-Council, i.e. at higher level of political career. WC 10 mentioned the value of time and the time limitations as a reason for her non-pursuance of this career in future. WC 2 has defined the political career in a different way. She said that she does not want to be directly in governance, but would like to work as a political leader so that she can work for the people. WC 30 mentioned the demand for Gorkhaland as the main cause for her pursuance of a political career.

The WCs of Darjiling know the meaning of a political career. However, they do not want to pursue it because of their domestic responsibilities, and to protect their values of womanhood in society. Thus, we observed in this case that the socialization process and consequent internalization of patriarchal norms are so deeply rooted in the minds of these women that they think that domestic responsibility is their priority in life, and for that they should sacrifice their career. A few of them were intelligent enough and gave politically correct statements, as they leave the decisions on their political career to the citizens.

6.3.1.1 Key Findings

1. Most of the women councillors have attained only primary level of education, but enjoy their economic freedom.
2. The physical mobility of the CIC councillors is higher than those of general councillors as CIC councillors are in the upper hierarchy of the political structure, and that is why they have more mobility. Most of the general WCs go beyond the boundary of the municipality, but their mobility is restricted within the district level.
3. There is no implementation of training for the councillors. This obstructs their social freedom.
4. Most of the WCs do not want to pursue a political career because of their family responsibilities, and a gendered notion of identity and performances.

6.3.2 Case of Balurghat

The social and economic opportunities, such as the average **educational** level among the WCs, are quite high in Balurghat. For example, four out of seven women councillors completed their graduation and studied beyond that. The remaining three have attained education up to primary level. The **occupational** empowerment among the women councillors is also very good, as four out of seven councillors are school teachers, beside three being housewives. It shows that a majority of them have their own economic freedom.

In this city's context, the physical **mobility** of the chairman is very high. She has visited Delhi, Hyderabad, Kerala and Chennai, beyond the state of West Bengal. In comparison, other WCs are less mobile, and they have only visited Kolkata. Their physical mobility is thus limited within the state of West Bengal. Therefore, the extent of physical mobility becomes higher following their positions in the political hierarchy. Thus, the physical mobility and position in administration are the most influential factors for empowerment of women. Kumar (2006) also said that empowerment of women in the administration is also a tool for economic, political, and social mobility for women.

About the availability of **training** facilities, the WCs said that there is an arrangement of training for the councillor, but it was not implemented. However, there was a discussion forum arranged in another city of south Bengal—Bardhaman—related to the responsibilities of the councillors. This forum helped them to understand their duties to some extent. This observation makes the importance of training visible in the empowerment of women. Gupta (2006) also suggested in his research the benefit of pressurizing political parties to make the leaders understand the significance of systematic training for women to generate empowerment.

The WCs' confidence to speak in public has been modified through the process of adaptation and experience. However, WCs feel shy and afraid when they speak for the first time in public. They acknowledged the role of the chairman in raising confidence among them. WC 10 said, 'The first time, I was really very afraid then Sucheta (the Chairman) from the back told me to speak and then I did'.

Except WC 15 and WC 21, all the WCs were engaged in politics during their student life in college. They also hail from families having political backgrounds. These two criteria influence the emancipation status of the WCs. Nair (2012) also stated that family is the primary training ground for one who aspires for a career in politics.

Support from the family is also an important indicator for the women who aspire for a career in politics. In Balurghat, unmarried councillors acknowledged the support they received from their mothers, while the married WCs were glad for the support of their husbands and fathers-in-law. One unmarried WC (WC 18) also expressed gratitude for her father, when she said, 'I do office work while my father does field work, i.e., visit to the ward'.

The above statement validates that the women themselves are confused and cannot differentiate between support and proxy representation. The statement of WC 18 seems to be an example of proxy for us, but she considers this as support. This is an ideal example of crucial point of contestation in the process of research where the theoretical knowledge of the researcher and the perceived knowledge out of the lived experiences of the research participants vary from each other.

Besides family support, **satisfaction from the position and gaining respect** are also linked with the aspirations of women representatives in the political field. Nanda (2012) in the study of Rajasthan also observed that public interaction has been noted to raise self-confidence among women representatives. In Balurghat, most WCs are satisfied with their position as well as with the respect they receive from the citizens. However, there are some different observations as well. For instance, WC 20 said that she gets respect from the people of her ward and she likes it. However, this respect sometimes takes a high toll as people would like to talk to her whenever she is out walking or doing something in public, and because of this, she loses her precious time. Because of this, she dislikes the cost she has to pay for this respect.

The women also question the respect, and they do not believe the people are actually sincere. WC 9 added in this context that, 'It is very difficult to believe the voice of local people as they say something in front and something else at the back'. In this context, WC 21 said:

> I have to consider the local people who are in front of me because people in behind always say bad things; there was a large population in front of Indira Gandhi, the only lady Prime Minister of India; however some people in back criticized her. She considered only the people who are in front as knew the humans have good as well as bad qualities also.

Therefore, in Balurghat, though WCs are satisfied with their position, they do not trust the citizens for whom they are serving. This distrust influences their leadership qualities as well. Being the grass-root-level representatives, they are less politicized in their behaviour, and as a result, sometimes this distrust of the WCs affects their behaviour towards their ward people.

The satisfaction they derive from the position and other social and political opportunities is directly linked with the aspiration of the WCs about their **vision of the future**. Most of them said that the pursuance of a political career depends either on the situation or on the party decision, and also the condition of their own physical and psychological health. However, most of them do not want to pursue a political career because of their family responsibilities. Even the WCs who were unmarried at the time of their election want to stop their political innings here, as they are now married and have huge responsibilities. WC 18 said that she wants to stop her political career because she has her own family to take care of. When she joined politics, she was single and, therefore, she did not have family responsibilities. Nair (2012) has also identified that family responsibility is an important reason for women to withdraw their name from politics. Therefore, family responsibilities become one of the major obstacles for the WCs, preventing them from going further in their political career and nominating themselves for the next tenure.

6.3.2.1 Key Findings

1. Educational level and economic freedom among the WCs are quite high. It also helps to increase their emancipation in their political career as well.
2. Except for the chairman, the physical mobility among the women councillors is limited within the state of West Bengal. This limited physical mobility influences their emancipation status in a negative way. Though the encouraging behaviour of the chairman acts as a positive factor to provide them with confidence, the political empowerment of women is relatively low because of other prohibiting factors.
3. WCs do not want to pursue their political career because of family responsibilities. Even unmarried WCs, who have the family responsibilities after their marriage, want to withdraw their name for the next election. Therefore, family responsibilities act as a negative factor to sustain their political career.

6.3.3 Case of Raniganj

In the case of **educational** level of the WCs in Raniganj, three are graduates, one studied up to high school, and the rest have attained primary level of education. Therefore, their educational level is also not high enough to attain higher economic freedom. Though in Darjiling we have observed that WCs with less educational qualification attain higher economic freedom, here the situation is very different. Thus, the location of a city, and the social and cultural contexts are very important in ascertaining the socio-economic freedom of the women councillors.

The economic status of the WCs of Raniganj is quite low, as can be determined by the **occupational** status of the councillors. The WCs are housewives, and they depend on their husbands in terms of economic freedom. Ghosh and Lama-Rewal (2005) said that unemployed housewives have more time on their hands for doing the work of local governance. However, this finding is not corroborated by the present study, as the time-use pattern reveals the fact that the housewives are very busy with the demands of domestic tasks.

The physical **mobility** among the WCs is highly variable. Most of them have visited places such as Bardhaman, Kolkata and Asansol for meeting purposes, but this travel is limited within the state of West Bengal.

In the context of **training** facilities, most of them have agreed that they did receive training about the role of councillors, which was held in *Nishan Bhawan* in Raniganj in the year 2010. The training was held two times. They agreed that the training is very good, and they learned important aspects about the responsibilities of the councillors. Similarly, in Maharashtra, Sekhson (2006) discussed about the role of *Alochana*, a non-profit research and documentation centre for women in training women. In Orissa, the City Managers' Association Orissa (2009) has stated that the government has developed a training kit for political education and experience for the women. This initiative taken by Orissa developed the knowledge among the women leaders about their responsibilities and about their capabilities in the meeting. Nair (2012) has also mentioned that such continuous trainings have a cumulative effect to empower women in governance. However, in our study in Raniganj, we observed that there is no compulsion to attend the training, so most of the WCs do not attend the training. They highlighted various reasons for this non-attendance. For example, WC 14 was not able to attend the training due to her daughter's education. WC 4 said that she does not attend any training, as she learnt her responsibilities from the ward secretary. WC 21 in this context said that there is training arranged at an interval, but she does not attend it as she has no information about the training. As she is from opposition party, ruling party members do not share the information of training with her. However, most of the WCs have received proper training about the role of the councillors, and have sufficient knowledge to implement their role. Gupta (2006) suggested that training is necessary not only during the period of women's entry into local government, but also in the post-entry period, for developing communication skills to enable them to function effectively.

WCs of Raniganj have acknowledged the **support they get from their family**. Everyone has given credit to their husbands. WC 12 said, 'My husband and son are very accommodating in nature and they do not say anything even if they do not have any food to eat'. WC 14 said, 'My husband took care of my little girl after I was elected as a councillor'. WC 1 acknowledges the support of her daughter-in-law as she does the household work for her. The daughter of WC 21 does the household work to give relief to her mother. Others have paid workers to do their household work. Thus in Raniganj, the use of paid workers is very common among the women councillors. However, in Darjiling and in Balurghat—i.e. in north Bengal—the concept of paid workers to provide household services is not at all common among the WCs.

The perception about the **pursuance of a political career** is more or less the same as the WCs of other cities. Family responsibility is an important factor for their non-pursuance of a political career. Most of them also added health problems as a reason for their non-pursuance. There were also some different observations. WC 1 wants to become the councillor again, even though she has to listen to some hard language and to harsh words. WC 13 said, 'I do my work but the pursuance of political career is not only my decision'. WC 12 proudly said that she wants to be a Member of the Legislative Assembly (MLA). However, she got the ticket of MLA in Jamuria, which is far away from her house, and that is why she did not take up the ticket for election. The distance became a significant deterrent factor for her because of two reasons. One is the insecurity of going to a distant place on her own, and the other is the domestic responsibility which she cannot perform well if she is going out somewhere far. Thus insecurity and domestic responsibility limit the scope of WCs' participation in governance.

Most of them are quite **satisfied** with the power **position** they have and the respect they get from the people of their ward. Again, there are some different observations as well. WC 21 is totally unsatisfied with the power and position she has got. The reason is that she is not able to do her work in her own way. She is elected from the opposition party, and the ruling party members obstruct her work.

Some women have a different voice and they observe this position as very temporary. WC 12 knows very well that she is elected under a reserved seat, and this seat will not be under reserved category in the next election. She thus narrated:

> I do not think that I have any power. After five years, this position will go. Many people have told me that 'didi you do not know who you are'. I said to them that I do not want to know about it. After five years, I will be a common person like you.

Beall (1996) also highlighted the positive abilities of women councillors to maintain good relations at local as well as neighbourhood level. Nanda (2012) stated that it will also help women to increase their level of self-confidence.

6.3.3.1 Key Findings

1. Most WCs do not want to pursue a political career because of a number of reasons such as health problems, social security and family responsibilities.
2. There is an arrangement for training the WCs. However, most of them did not attend as there is no compulsion to attend the training in their job rules. Therefore, legal compulsion is necessary to encourage women representatives to attend the training.
3. There is information blockage based on party affiliation. WCs elected from the opposition party do not get the information about training facilities. This is also an obstacle to increasing the emancipation status of women.

6.3.4 Case of Hugli-Chinsurah

The **educational** level among the WCs is quite high here. Among the WCs, three are graduates, four studied till higher secondary level, and two have completed their primary education. However, **economic** freedom is very low among them. They are housewives, and do the work in their household. They are fully dependent on their husband's income. Here this observation differs from that of Darjiling, where we found that women WCs are economically independent even though they have lower educational qualification.

Kodoth and Mishra (2011) have stated that women are constrained by restrictions on their mobility because of moral policing and less disposable time owing to their household responsibilities. In Hugli-Chinsurah, physical **mobility** is very low among WCs. They are not even mobile within the city, as their husbands do the outdoor work. They have visited Kolkata once as councillors for the purpose of training. WC 20 said that there is no space for practical implementation of the subjects they have learned in training. Only WC 17 said that she has visited Kolkata another eight times to attend the training. The municipality recommended her name every time for **training** as other WCs did not want to go. A similar observation is also found in Hugli-Chinsurah if we compare the situation with Raniganj. As there is no compulsion in case of attending training, WCs do not prefer to attend the training. Another reason is that since most of the WCs never do the outworks within the municipality, they do not prefer to go to distant places to attend the training. In case of efficiency building after training, WC 17 added:

> Training is good. I learned many unknown aspects of governance after training. However, implementation depends on the councillor as other WCs remain non-vocal in the known aspect even after they learned it in the training.

To overcome this problem in Tamil Nadu, the Women Leaders Federation was developed for building capacity among women leaders to manage crowds and the media (Palanthurai, 2012). It helps to make women leaders more visible. Mumbai government has also taken an initiative of leadership training in collaboration with the Netherlands to make women transformative leaders (Holzner and Wit 2003: 1–2).

Except WC 17, all of them have got **inspiration from their family** to come into the political field. The family encourages them to secure the seat for the men of the next generation. In this context, it is important to mention as stated by Ghosh and Lama-Rewal (2005) that family is a residual resource, besides the resources such as time, education and economic self-reliance. However, in case of Hugli-Chinsurah, family inspiration acts negatively for the WCs as it encourages proxy nomination of women's representatives, and this is clear from the statement of a few WCs. WC 25 said that, 'Political party had decided to nominate my husband for the councillor's position. However, when the seat got reserved for women they decided to nominate me'. WC 2 confessed, 'Political party gave offer to my husband first, but my family forced me to contest for the position when the seat of this ward got reserved for women'.

According to Rizvi (2006), such forced participation of women ultimately hampers their capacity of decision-making as it depends on the permutations and combinations of the influential family members. However there are exceptions to this. WC17 gets prepared food delivered to her house from her mother to reduce her household duties. In the morning before going to duty, her husband helps her to reach the meeting place where all the workers meet and do ward development work, and she does her inspection there.

However, in case of other WCs as stated earlier, the proxy is very visible. The family supports them to do their work, not by means of doing household chores, but by doing the work of the councillor. In this context, Rizvi (2006) observed that family members give no concession in domestic chores, and thus social norms and taboos adversely affect the participation of women in the affairs of urban governance. Except for WC 17, all the WCs have divided their work into two groups, that is, indoor and outdoor work. In the case of outdoor work such as ward visits and giving time to the party office, men of the family or associated party members do the work. In the case of indoor work, WCs meet the people at their home and give out certificates and other help in between their household work. This phenomenon has also been identified by Bari (2005), who stated that the patriarchy behind it has placed the women within the private arena of their home even if they are in governance.

Women councillors have to face a number of other problems linked to the social norms as mentioned by WC 27:

> People come to my house all the time. I wear causal dresses in house and sometimes feel uneasy. As I do not wear sari, people who come to my house look at me curiously.

This is a negative aspect of the social norms which women have to follow. The women are victimized because of their dresses, as they wear casual clothes at home, which is unfit to go out in public. If they meet the ward people in such dresses, people pass comments, but if they take the time to change into proper dress, people are too impatient to wait for them to do so. Omvedt (2005) identified cultural patterns as a reason for the striking difference in dress and in behavioural patterns between men and women, and in this difference, women suffer the most.

Though most of them come from the political family background, they have no clear vision for their future. They opined that it is the decision of their concerned party whether to nominate them or not for next time. In this context, Ghosh (2003) stated that the re-nomination of the women candidates depends on the political equation of the concerned party rather than on their capability and acceptability to the common people. In the urban setting, Nair (2012) identified that such a rotation system made the future of the women councillors uncertain. However, in our study in Hugli-Chinsurah, six of them want to pursue a political career as a councillor to work for the society. WC 27 said that she wants to pursue her career as she wants to complete the unfinished work in her ward if she gets a chance for another 5 years. WC 17 wants to work for the people from the ground level as she thinks higher positions bring higher problems. WC 23 wants to pursue her political career, but she has argued that though she is experienced in the political line, she has never got a good position as a senior. Therefore, there is discrimination against the women members within the political party. Though she is an experienced member in the political field, she never gets CIC position.

In the context of **gaining respect** from ward people, most of them have confessed that they got their popularity and respect because of the popularity of their husband or brother-in-law. WC 27 said:

> Ward people respect me because they knew my husband from his childhood. He behaves well to all the people and is popular as an All India Trinamool Congress party member. However, when I was elected as a Councillor, some people of my ward tried to do some illegal works with my permission as they thought I do not know about them, but my husband informed me about them.

Therefore, knowledge of the husband about the ward sometimes has the positive outcome of providing information about the ward people. WC 17 said that she is like a daughter or a daughter-in-law for the people of her ward. WC 23 treats her ward people as her family. In this context, WC 16 showed enormous confidence about her identity in the ward as she could calculate that among 10,000 voters, 95 per cent will vote for her surely during the next election. Most of them have a positive attitude about their citizens.

6.3.4.1 Key Findings

1. The educational level of the WCs is very high. However, economic freedom is low.
2. Family inspiration acts negatively for the WCs as it encourages proxy nomination of women's representatives.
3. Their physical mobility is also low as their outdoor works are done by the men of their family and political party. They have no clear vision about the pursuance of a political career as it depends on the decision of their political party as well as family members.
4. There is discrimination of the women members within the political party. Despite being experienced members in the political field, WCs never get CIC position.
5. Like in Raniganj, as they have no legal compulsion, they do not prefer to avail the training facilities because of family responsibilities as well as distance of the training venue. However, all of them availed the training only one time.

6.3.5 Overall Findings

In case of socio-economic freedom, Darjiling scored negatively on the basis of composite scores (Table 6.6). Most of the WCs in Darjiling have studied up to the primary level. However, they enjoy higher economic freedom. They have no opportunity to get training. In Hugli-Chinsurah, the socio-economic freedom of the WCs is high as they have higher educational qualification and other social opportunities. Economically they are not independent. Further, due to proxy nomination of the WCs, the political freedom among them is very low. This signifies that they have the capability, but do not have space for implementation. The composite score values of Raniganj and Balurghat run near to zero, which signifies the medium status of socio-economic freedom (Table 6.6).

In the context of socio-economic freedom of women representatives, Srivastava (2015) opined that education and reservation are not enough for women to join governance and politics. They need a supportive family and a family that has a political background, someone in the family who is willing to take over the care of the family and children and a good amount of surplus money. In case of money for women, Srivastava (2015) also suggested that microfinance can be a good option to make the elected women representatives economically independent.

Table 6.6 Socio-economic freedom of the women councillors

Women councillors	Indicators of socio-economic freedom	Value	Darjiling	Balurghat	Raniganj	Hugli-Chinsurah	Mean	Standard deviation
	Have education HS and above	Percentage	42.86	57.14	50	77.77	56.94	15.059
		Z score	0.935	0.013	0.461	1.383	–	–
	Have physical mobility beyond municipality	Percentage	7.14	71.43	87.5	88.88	63.73	38.55
		Z score	1.468	0.199	0.616	0.652	–	–
	Have training	Percentage	0	28.57	87.5	88.88	51.23	44.23
		Z score	1.158	0.512	0.82	0.851	–	–
	Wants to progress in their political career	Percentage	42.86	42.86	62.5	66.66	53.72	12.65
		Z score	0.858	0.858	0.694	1.023	–	–
	Occupational status	Percentage	28.57	42.86	0	0	17.85	21.42
		Z score	0.499	1.166	0.833	0.833	–	–
	Composite score of socio-economic freedom		−3.920	0.008	0.836	3.076	–	–

Source: Field Survey, 2013–2015

6.4 Transparency and Protective Security of the Women Councillors

According to Joseph (2011), transparency means the openness of the people to deal with one another with full disclosure and lucidity. The openness of the people depends on the trust towards other people. Joseph (2011) identified the following indicators of transparency: access to information, active role of media and corruption.

Joseph (2011) also said in the context of protective security that it is needed to provide social safety. It is a matter of institutional arrangements. Women face many problems in politics as they come from different social segments. They need social support to cope with problems and perform well in this field. Among the problems, violence is used as a tool to restrict a person's entry into the political field. However, Upadhyay and Bhardwaj (2006) observed in their study that though patriarchal society does not like women in politics, there are no cases of violence against women in the urban local bodies. They have also identified the reasons for this, since a majority of elected women come from political family background or have party backing, which makes the chances of violence against women minimal.

6.4.1 Transparency

Transparency includes access to information such as newspapers, government reports, media coverage and awareness about the prevalence of corruption, as identified by Joseph (2011). In **Darjiling,** there is an information blockage related to the distribution of newspapers and reports among the women councillors, based on their position in the political hierarchy. CIC councillors have access to newspaper and reports. However, general councillors have no access to newspapers and reports. This kind of information blockage is a major obstacle for women to participate in the governance as stated by Palanthurai et al. (2009).

Khattak (1996) identified the media and the active role it plays as an important indicator of women's political participation. In Darjiling, eight WCs came into the purview of the media. However, two CIC councillors said they dislike the publicity. Among them, one WC shared her experience with media on the incident of a cloudburst and consequent landslide. A Bengali reporter had come to take an interview about her role in managing this calamity. This has also been published in a Bengali newspaper which is, unfortunately, illegible to her as she cannot read Bengali and the news came in Bengali only. That incident made her disappointed with the media because it did not take care of her own language, and did not provide any translation script for her. Therefore, a common communicating language is necessary to make media more effective to the WCs of a particular region. Other WCs do not like publicity, and that is why they never share their experiences in front of media.

Though they do not like to share their experiences, they do have a lot of knowledge about the inherent system of the municipality. In case of corruption, WC13 said in a devastating manner that she does not know about monetary usage and tax-related information. She has also mentioned that the staff is highly corrupt and takes bribes from people. In the opinion of WC 17, the low salary is indirect encouragement for the councillors to take bribes from people. Therefore, it is observed from the present study that WCs are against bribes, and this is also proven from their confession about corruption. A similar observation is found by Beaman et al. (2006) in their study where they observed that women representatives are good policymakers, and they take action against bribes.

In **Balurghat**, the experience of first-time exposure to media was quite frightening to the WCs. After that, it has become more familiar to them. For example, WC 17 represents herself as a protester in front of the media on the issue of unemployment. She added, 'I have told whatever the truth is and have always tried to represent what I am'. As a result, it brings transparency to the sphere of governance.

However, there is no circulation of newspapers and government reports in the municipality. This leads to a blockage of information among them.

In **Raniganj,** most of the WCs said that they can access newspapers and government reports in the municipality. However, these are provided by the political party. In case of media coverage, except for three WCs, all of them represent themselves in front of the media. Among the three, two WCs (WC 14 and WC 17) expressed their fear to come in front of media, and WC 21 never came in the purview of media, but because of a different reason as she expressed:

> Media comes to me but if I say my own feelings then my party members will lose their popularity. For that reason, I told the members of media to go to my ward secretary to know about the development of the ward.

From this argument, it is clear that she knows many untold facts about the members of her concerned party, but she is cautious about telling those in public. That is why she requested the media to interview the ward secretary for any kind of news regarding ward development. This is the silent process of changing the focus from WC to ward secretary. Shah (2015) observed a similar process in Pakistan, where the leading women MNAs have a fear that they would be sidelined at any point without any satisfactory reason given by the male leaders of the concerned party.

In **Hugli-Chinsurah**, it is observed that the WCs, who are elected in proxy seats of their men family members, come into the purview of media only once. This is because the men of their family handle the media exposure.

WC 17, who is elected in the reserved seat of scheduled caste but not in proxy, has said that her interview was published in the Bengali newspaper named *Bartaman*. After that, local channels have interviewed her to learn more about the problems of the waterlogged area in her ward. Due to geological factors, this area remains waterlogged and prone to floods all the time. For this reason, during the flood, while she was working in that zone to save the people of her ward, she gave the interview.

On the other hand, WC 20 hates to come before the media as her ward is discriminated against in terms of ward development. She is elected from the opposition party, and the media always asks about the development of her ward. She feels uneasy in front of the media in such an adverse situation. Therefore, not only proxy nomination but also party-based opposition are influential factors for the non-exposure to media and disempowerment of WCs.

6.4.2 Protective Security

In **Darjiling**, official staff and men councillors (MCs) do not impose a gendered outlook for WCs, and WCs also do not prefer to take favour from them for being women. However, though the WCs are empowered enough, they face certain internal problems among them based on language. This language barrier sometimes acts as a cultural and social barrier as well. WC 17, who speaks in Bengali and originally came from the plains, always stays separate from the rest of the WCs (who are

Nepali-speaking people). All others also maintain their distance from her. The reason is that she speaks in Bengali, and she is from the plain region of West Bengal. There is a long history of Gorkhaland movement which has been elaborated upon in Chap. 2. This historical factor causes people of the hilly region of Darjiling to have strong sentiments about their regional and communal identity. This regional identity is a strong reason why WC 17 faces this distance. In this context, WC 17 said:

> Government of West Bengal is afraid of *Nari Morcha* (a group of women from hills who are fighting for Gorkhaland) of the hill. However, there is no leader among them due to problems of ego and selfishness among them. I belong to plains, and I am educated, that is why, other WCs are jealous.

Language is a medium of communication. However, here it creates division among the WCs, as one of them speaks in Bengali and the rest are from the Nepali community. Though the medium is not a barrier, as they can use their common language Hindi, there is a large psychological barrier due to such cultural differences. Though the Bengali-speaking WC is in Chairman-in-Council position, she remains isolated because of miscommunication with others. It makes the WCs more vulnerable due to lack of networking among them.

Councillors always discuss their own decisions with other councillors sitting in the common room. The chairman is always accessible to them. As for membership in other social organizations, only one of them has membership to the Red Cross society and voluntary youth club organization.

In **Balurghat** WCs get maximum support from their chairperson. As she is a woman and also a political struggler, she can understand the problems of WCs better. As a result, WCs also acknowledge the support they get from their chairperson. Even on days when they have to return home late due to extra work in the municipality, the chairperson arranges a car to drop the WCs. She always provides encouragement to the WCs to empower them. While explaining her responsibilities, the chairperson argued:

> I was a Chairman for many years. However, the word 'woman' has always placed me behind. I have faced and overcome many obstacles as a woman Chairman. However, still I am not free but will be free when my identity becomes only Chairman rather than woman Chairman.

Her argument clarifies that despite being in the post of chairman, she is also struggling for her identity as a woman.

Besides the encouragement of the chairperson, WCs are also attached with other social organizations. This leads to increase in their exposure levels as well. WC 10 said that she was an athletic champion and was the president of the 'women's athletics organization'. She is also a member of an eye donation organization named 'Prayoso'. WC 12 as a primary teacher has membership of 'Sara Bangla Primary Teacher Samiti'. The chairperson is also a member of the 'district sports association' and an executive member of the 'Red Cross organization'. WC 18 is a member of the social work group and blood bank.

In Balurghat, it is found that WCs enjoy their political, social as well as economic freedom, as they have a strong base of educational attainment as well as political experience. It is also evident that due to their unmarried status, most of them are free from the extra burden of family chores. The result is the ultimate engagement in the field of governance. Another pull factor is the encouragement of the woman chairperson, which helps the WCs to attain more visibility.

In **Raniganj,** the division among the WCs based on their political party affiliation is very prominent. WC 9, elected from the opposition party, argued that she does not have close relations with anyone in the councillors' circle because of the mismatch of thoughts with others. However, she does not think of the other WCs as her enemies. She feels that the other WCs do not know their power and remain like slaves of their political party. She filed a complaint against WC 13 (elected from ruling political party) as she closed the only health centre in her ward forcibly, though it is very much needed. She requested many times but nobody listened to her request as she is from the opposition party. Ultimately, she managed a doctor on her own to solve the problem. Besides this political turmoil, this fact reveals the concerns of the WC towards public health of her ward. Similar observation of women's entrepreneurship in doing their own work has also been cited by Doepke et al. (2012). They have stated that when women are involved in governance, the composition of spending shifts towards health and children. Beall (1996) said that women's social responsibility towards their home also includes family health care, and that is why when they are in governance they give more priority to health.

Most of the WCs of the ruling party acknowledged the support of the chairman. For example, WC 1 said that in any circumstance, he is always supportive and he always gives respect while talking and also listens to their words with patience. WC 12 added that the chairman does not say anything harsh, even if she has an outburst at times. According to WC 14, he is a very accommodating person, and in the situations when opponents are more active, he is the only support to them.

However, the WC from opposition party (WC 9) thinks that the chairman is not good. She also added that it does not matter whether the chairman is man or woman; it is party colour which is important more than anything else. WC 21, elected from a subgroup of the ruling political party, is also upset with the behaviour of the chairman. She said that he does not care about whatever she says about the development of her wards. He only says some irrelevant words to console her such as *dekchi* (means 'I see') and *vabchi* (means 'I will think about it'). He gives priority to the voice of the ward development secretary, not her.

Among the WCs, no one has membership to any social organization. Only WC 9 has organized an 'association of Mazdur' (labour organization) in her ward.

In **Hugli-Chinsurah**, WCs maintain good interpersonal relations but this is mostly superficial. According to Palanthurai (2012), this kind of lack in networking among the women representatives makes them more vulnerable to the hidden patriarchy in the governance system. They have no common voice about their demands, which is why it remains unimplemented in most cases. In the context of the chairman's support, all the WCs agreed that the chairman takes their arguments casually, and he does not implement anything which they have said. From this argument, it is clear that most of the WCs are elected in proxy seats of their male family members, and that is why the chairman does not consider their arguments as important. As a result, they remain proxy not only in their work and decision-making abilities but also in placing arguments. Such phenomena make the WCs more non-vocal.

6.4.3 Overall Findings

Among the indicators of transparency and protective security, only two indicators from two ends can be analyzed quantitatively, while the other indicators are better explained through the qualitative methods in the previous section. From the qualitative explanation, it has been found that though political freedom of the WCs of Darjiling is higher, the level of transparency in the municipal system to provide them with exposure is very less. In Balurghat, WCs have a friendly environment as well as good media exposure. Moreover, WCs are also engaged in different social organizations to provide service to the society beyond their political arena. In Raniganj, party-based opposition and controlling nature of the ruling party obstruct their voice in front of the media as well as in public. In Hugli-Chinsurah, proxy nomination is the major cause of discrimination in the context of media exposure. They do not even have membership of social organizations (Table. 6.7). This is reflected in the quantitative analysis as well. Therefore, from these observations, it is clear that reservation is not the only a way to empower women representatives, as there are other significant factors affecting the situation. These should be considered while taking any measures regarding empowerment of women's representatives in the government.

Table 6.7 Transparency and protective security of the women councillors

	Transparency		Protective security		
Cities	% of women councillors who come under the purview of media (A)	Z scores	% of women councillors who have membership in social organization (B)	Z scores	Composite score of transparency and protective security (A + B)
Darjiling	21.43	−1.49	14.29	−0.77	−2.26
Raniganj	75	0.51	12.5	−0.89	−0.38
Balurghat	71.43	0.37	42.85	1.16	1.53
Hugli-Chinsurah	77.77	0.61	33.33	0.51	1.12
Mean	61.41		25.74		
Standard deviation	26.78		14.79		

Source: Field Survey, 2013–2015

6.5 Functioning of the Women Councillors

6.5.1 Decisions Related to Developmental Work

WCs (women councillors) of the cities perform different functions in their area of concern. In all cities, these are of two kinds of WCs—the CIC councillors and general councillors. CIC members are the ones who hold responsibilities of different departments such as water, land, basic services, poverty eradication, etc. Beside their own wards, they have to look into departmental problems as well. They are also in a higher position in the political hierarchy. The general councillors have responsibilities only of their own ward development.

In **Darjiling,** WCs deal with the issue of the development of basic services such as water, electricity, road, sanitation, slum area housing development, sewage, garbage removal, development of *jhora* and *nala* (drains), disease protection, Below Poverty Level cards, earthquake remedies, implementation of government schemes, giving certificates to the residents, birth and death certificates, stopping family squabbles, etc. They give proposals linked to all sorts of functions, as mentioned earlier to the chairman in the meetings.

WC 14 has developed an innovative idea of categorizing the people of her ward for providing funds to them. It has facilitated smooth functioning of the work of fund distribution. She constructed a community hall for meetings in her ward from the funds of the Member of the Legislative Assembly (1.5 lakhs) and MP (5 lakhs) out of their area development quota. She also stated that 'it is our area and our problems; therefore we have to solve them'. Kumar (2006) says that implementation of income-generating programmes will lead to the development of basic services in urban areas by women administrators. WC 29 of the water department has enhanced the work culture in her department. She improved the condition of water supply in the municipality by checking the illegal supply of water. Similar observation has been found in the study of Beaman et al. (2006) and Kulkarni (2011). They stated that women representatives as water managers perform better on the management of water, and they give more priority to drinking water facilities. Upadhyay and Bhardwaj (2006) also argued that women understand the difficulties of not getting proper water supply—since women have to fetch water if the water is not available, so if they are in the administration, they will make full efforts for the betterment of basic services.

In **Balurghat**, WCs deal with issues such as development of roads, drains, light, tap water connection, electric pole, latrine, podium, sanitation, cleaning of garbage and housing development, etc. However, their main concern is to improve the potable drinking water supply in the city. According to the study by Gupta (2006), women representatives are more sensitive to urban management issues and problems. However, in the present study, we have observed that in a few cases, they depend on men such as their father, husband, or ward secretary for outdoor works. The chairperson as a woman holds the responsibilities of Balurghat city as a whole, and she performs very well as she is at the top of the hierarchy, and does not have to follow the dictates of men as boss.

In **Raniganj,** WCs provide basic services to poor. However, the main concern of Raniganj is to improve drinking water facilities and to generate employment, as most of the people are unemployed due to strikes and closure of industries in the area. To eradicate the problem of water supply, WC 9 has developed timely connection of drinking water in her ward. To help poor families, WC 1 helps the Below Poverty Level families to get free rations. WC 1 also provides employment to the people under different government schemes in her ward.

In **Hugli-Chinsurah**, though most of the WCs are elected in proxy seats, they are concerned about their ward development issues. The main concern of the WCs is to focus on the infrastructural development and flood management as the city is located on the banks of the Bhagirathi River and often gets flooded. Water supply is not an issue here like in Darjiling, Balurghat and Raniganj. WC 27 and WC 9 said that they sanctioned a park for children in their wards. WC 17 has developed water connections and proper electrification of all roads of her ward and also in the boundary areas. To reduce the frequency of flood in the waterlogged areas of her ward, she made a high drain. WC 25 has developed an innovative strategy for the inspection of the development of her wards. She has divided her ward into nine smaller neighbourhoods, and has selected a representative from each locality to get better information about the development of that locality.

From the previous discussions, it has been found that the WCs provide services for the development of their cities. However, the major concerns of the cities are different. This is because of the location of cities in different geographical regions of West Bengal. For example, in Darjiling, the supply of water is important; in Balurghat the main concern is to provide potable drinking water; in Raniganj it is unemployment; and in Hugli-Chinsurah the focus is on infrastructural development. A similar study of Chattopadhyay and Duflo (October 2001, revised 2003) describes how the concerns of women representatives vary with space. For instance, women representatives of West Bengal place more emphasis on drinking water facilities, whereas in Rajasthan, the priority is on road development. Beall (1996) argued that women's responsibility for household and community management gives them a particular vested interest in safe water supply, efficient sanitation, solid waste removal, and the provision of affordable and reliable sources of energy.

6.5.2 Women's Empowerment Programmes

Kodoth and Mishra (2011) found that women councillors are more courageous while working, especially on the issue of women's development and empowerment. However, in **Darjiling,** there are no women's empowerment programmes. According to the WCs, as there is no allotment of funds for this kind of programme, they are helpless to do anything on this front. There are programmes such as financing self-help groups, widowed and old-age pension, development of beauty parlours and Integrated Child Development Services (ICDS). These programmes have been stopped due to lack of funds.

In **Balurghat**, WCs undertake programmes related to women's empowerment, for example, development of self-help group and *Mahila Samiti*, employment generation for women through Sahbhagi Shikshan Kendra (SSK), ICDS, small-scale industry-related training programmes organized in *Matri Sadan*, and giving of sewing machines to women. WC 21 protested against a rape with the help of the women of her ward. WC 6 has also joined a national organization of *Mahila Samiti* in Delhi, and she was the president of Balurghat zone. Her organization works for women's empowerment, and getting more women to register in their organizations.

Joseph (2011) also stated that elected women representatives help to raise women's issues in public discussions as well as attack the social norms. Iyer et al.(2011) also stated that women's inclusion in politics will ultimately give a greater voice to women. However, Ghosh (2003) has made a different observation in her study. According to her, women representatives do not want to represent themselves as the leaders of women.

In **Raniganj,** WCs have taken up various programmes related to women's empowerment, for example, development of self-help groups and providing loans to them, giving work to the women in the ICDS centre under the Mid-day Meal Scheme and in computer coaching centres, providing grants to widows and old women and pension to widows, etc. WC 13 has developed a taxi stand which is supervised by women self-help groups. In this context, Gupta (2006) stated that women are also more approachable by other women and other vulnerable groups.

According to Doepke et al. (2012), when women are involved in politics as policymakers and voters, the composition of government spending shifts towards higher expenditure related to health and children. In this context, Gupta (2006) argued that most of the women councillors do not know much about their work in their first tenure, and that is why they deal primarily with the issues related to women. In **Hugli-Chinsurah**, WCs have taken various women's empowerment programmes. They arranged the courses of beautician, tailoring, learning software and hardware and formation of self-help groups, and appointed women of the ward as cooks under the Mid-day Meal Scheme. For women's empowerment, WC 17 joined the women's self-help groups as a member, not as president. Though she is eligible to become the president, she chooses to be a member, mainly in order to put pressure on the higher authority for any work related to women. WC 16 has developed a self-help group named *Bandhan,* which provides loans to women for training and other purposes. These initiatives of WCs indicate the improvement of gender sensibility at grass-root level, although we do not have much priority of gender-responsive budgeting, as stated by Haug (2012) in Nepal.

6.5.3 Attending Meetings and Visits to the Ward

In **Darjiling,** the pattern of attending meeting and visiting wards by the WCs varies from person to person. For example, WC 1 goes to visit the ward on the days when she does not go to the municipality. WC 29 attends office daily. She goes to the ward in the morning regularly, and stays in office in the afternoon. She also visits the ward and the zone of water sources at night with her local members, to prevent illegal supply of water. It is very important for a councillor to visit the ward to know the problems of their respective areas. For example, Paradza et al. (2010) in their study of South Africa stated that the level and frequencies of women councillors making contact with the residents and citizens are important, and that is why these visits can also be tracked by authorities to know about the councillors' work and decisions.

Attending office is important for the councillors to get knowledge about the schemes and other information, and also to contact the people of their ward. However, there is no formal rule of attending office for the councillors. That is why those who come from distant places do not attend the municipality daily. WC 2 states, 'I come from very long distance and pay fare of INR 40 per day. However, the women councillors who also come from a longer distance and do not come to office regularly encourage me not to come to the office'. Therefore, distance and additional cost of transport are matters that discourage the WCs to attend the office daily.

Beaman et al. (2006) says that women are less effective and receive bad responses in meetings as the meetings are mainly guided by men leaders. In Darjiling, this kind of experience is not cited by WCs, but most of them cited distance as an important factor for not attending meeting of the WCs. The meeting of Chairman-in-Council is held once a month, and the meeting of the board of councillors is held after an interval of 3 months. All the councillors are invited to the meeting through official letters dispatched from the municipality. For example, WC 2 has added distance as a factor for her low attendance in the meetings. As she comes from Ghum station (a place far away from the municipality), she faces several hurdles in her journey, which also takes her a long time. WC 1 and WC 2 come from 'Jalapahar' far away from the municipality and pay a high fare daily. However, all of them try to come on the meeting days, as without attending the meeting and putting their signatures on the official letter of meeting, they cannot receive their monthly payment.

In **Balurghat,** the majority of the WCs perform their work very well except for three exceptions. However, the father of the young WC 23 undertakes the ward visits every day, and the ward secretary of WC 3 does the ward visits and other work, on behalf of the woman councillor. The husband of WC 21 is also very popular in her ward and does the outdoor work. The other seven women councillors conduct their ward visits on their own and very regularly.

In case of attending meetings and placing forth decisions in the meeting, the woman chairperson encourages them. WCs have regular attendance in the municipality. On the meeting days and on the days of excessive work, the chairperson provides her own car to drop the women councillors to their house. The solidarity and caring attitude of the woman chairperson have helped the other WCs perform well in Balurghat.

In **Raniganj,** the performance of WCs is controlled by two factors. The performances of women from the ruling party are controlled by their party's male members, and the women from the opposition party are not allowed to perform by the ruling party. In this context, WC 12 said that she did not go outside her house during the first few months of her journey as a councillor because of fear as she had never gone out in public before. However, at present, she visits the ward daily. WC 21 divides her time for the work. For example, on the day of attending municipal office, she does not go to the ward for the visit and vice versa. WC 14 said that she visits the houses of the local people. WC 13 has said that she knows all the residents of her ward and she also used to visit the wards adjacent to her ward. WC 9 visits the ward especially to the lower-income group people and tries to remove their poverty. She also added that her main mission is to improve the conditions of slums with the help of government schemes.

In **Hugli-Chinsurah**, WCs attend the meeting as without attending it, no one gets their honorarium. They are informed about the meeting through an official letter or by a phone call in an emergency. Except for the meeting, all the relevant work at the municipality is done by the party members, ward secretary and husband of the WCs because of their proxy nomination. WCs are elected in the proxy seats of the men of their family and of male members of the party, to secure the seats for them for the next tenure. That is why, male members of their family and political party exercise the actual influence to manage the work of the councillor.

There are four groups of councillors in terms of their pattern of ward visits. One group of WCs, who are elected as the proxy of their husbands, give the responsibilities of ward visit to their husband. WC 23 said that, 'My husband goes to visit the slum areas as they use slang language while talking. It is an obstacle for communication with them for me'. This statement indicates that the male chauvinism and the associated language can be a barrier to women's work even at the local level. In the second group, we have the WCs who are elected in the proxy seats of the men of their family other than husbands. Here, the concerned male members conduct ward visits to reduce the work burden of the WCs as well as to increase their popularity in the ward. It will also help the male members to contest elections in the next tenure when the seat becomes unreserved. In the third group, male party members do the work of the councillor including ward visits on behalf of the WC. Thus, WC 9 said that:

> Municipality is near my house. However, I have no need to go there because my ward committee secretary does all the works of the councillor and also does the ward visits. Only in case of emergency, I go there.

In the fourth group, WCs do the ward visits and official work on their own. In this context, WC 25 has developed a similar strategy for the inspection of the development of her wards, as we observed before in case of decisions related to developmental work of the women councillors of Hugli-Chinsurah. She has divided her ward into nine smaller neighbourhoods, and has selected a representative from each neighbourhood to get better information about the development of that particular area. Stating this kind of observation in their study, Narain and Mridubhashini (2006) stated that there is no doubt that women do have their inherent weakness due to dual responsibilities both at home and their work place. However, it does not mean they are irresponsible and have no creativity in their work.

6.6 Conclusion

From the present chapter, the observations are as follows: political and socio-economic freedom of the WCs depends on the sociocultural set-up and political environment of that particular city. Transparency and protective security signify the institutional behaviour and psychology of other councillors. All these factors are indirectly influenced by the functioning of the WCs, and their achievements towards substantial freedom (i.e., implementation for the betterment of the society).

In Darjiling, people help to elect more women in urban governance due to positive psychology of the people towards women, and less gender gap in their society. However, the lack of networking among the WCs, the gap between CIC councillors and general councillors, and the conflict between Nepali and non-Nepali WCs decrease their functioning ability in the wards. Beside these, the WCs in Darjiling have done a lot of work for their wards, and got greater acceptance in their wards. However, there is no women's empowerment programme in Darjiling, though there are 15 elected WCs from 15 wards in the municipality. The reason is that they have very little knowledge about these programmes, as their education and level of social freedom are very low. Moreover, they face lack of availability of funds. In Balurghat, the socio-economic freedom of the women councillors is high. The influence of the woman chairperson with long political experience from her student life helps to bring confidence among them. She also provides a favourable environment in the municipality to provide transparency and protective security to them. As a result, functioning of the women councillors is also good, and they are aware of the importance of implementing women's empowerment programmes in their city.

However, in the cities of south Bengal, such as Raniganj and Hugli-Chinsurah, the freedom and functioning behaviour of the WCs are totally different from the cities of north Bengal. Inherent patriarchy in terms of opposition and proxy representation is very prominent here. However, despite these obstacles, some WCs struggle to achieve freedom to perform their function in society.

Therefore, any generalization regarding freedom and functioning behaviour is not possible. It varies not only with space and time, but also with individual capability level of the WCs (Samanta and Banerjee, 2016). However, the explanation of the present chapter helps to understand the capability of the WCs in the cities of West Bengal.

References

Bari F (2005) Women's political participation: issues and challenges. Draft. United Nations, Division for the Advancement of Women (DAW), Bangkok, Thailand. www.un.org/womenwatch/daw/.../EGM-WPD-EE-2005-EP.12%20%20draft%20F.pdf. Accessed 13 Dec 2012

Beall J (1996) Urban governance: why gender matters? https://www.gdrc.org/u-gov/doc-whygendermatters.html. Accessed 30 Jan 2014

Beaman L, Duflo E, Pande R, Topalova P (2006) Women politicians, gender bias and policy making in rural India. State of the World's Children 2007, Background Paper, UNICEF, pp 1–23. https://www.unicef.org/sowc07/docs/beaman_duflo_pande_topalova.pdf. Accessed 20 June 2013

Bharathy M (2012) Gendered citizenship for deepening local democracy: a study of ten villages in Maharastra. In: Palanthurai G (ed) Elected women representatives and empowerment. Concept Publishing Company, New Delhi, pp 79–96

Chattopadhyay R, Duflo E (October 2001, Revised 2003) Women as policy makers: evidence from a India-wide randomized policy experiment, pp 1–21. weber.ucsd.edu/~tkousser/WomenasPolicyMakers.pdf. Accessed 20 June 2013

CMAO (City Managers' Association Orissa) (2009) Local government management learning kit for elected women representatives of urban local bodies of Orissa. Housing and Urban Development Department, Government of Orissa, Orissa. www.cmao.nic.in/2009/Events/LGM%20Women%20Orissa%20Module.pdf. Accessed 26 Nov 2013

Doepke M, Miche'le T, Voena A (2012) The economics and politics of women's rights. Annu Rev Econ 4(6):6.1–6.34. https://doi.org/10.1146/annurev-economics-061109-080201

Garrett DJ (2005) Amartya Sen's ethics of substantial freedom, pp 1–7. http://people.wku.edu/jan.garrett/ethics/senethic.htm. Accessed 1 May 2016

Ghosh A (2003) Women's reservation and electoral politics in urban local bodies: an analysis of Chennai municipal corporation elections. Indian J Gend Stud 10(1):118–140. https://doi.org/10.1177/097152150301000106

Ghosh A, Lama-Rewal ST (2005) Democratization in progress (women and local politics in urban India). Tulika Books, New Delhi

Gupta S (2006) Political empowerment of women in urban local bodies and the 74th Constitution amendment: dispelling doubts and devising future strategies. In: Singh UB (ed) Empowerment of women in urban administration (experiences and strategies). Serial Publications, New Delhi, pp 67–80

Haug M (2012) Women's political participation at the sub-national level in Nepal: a mapping study (UN WOMEN). South Asia Country review series on women's political leadership. Norwegian Institute for Urban and Regional Research; UN Women, Norway, pp 5–20. https://asiapacific.unwomen.org/en/digital-library/publications/2012/6/womens-political-participation-at-the-subnational-level-in-nepal-a-mapping-study. Accessed 30 Apr 2013

Holzner M, Wit JD (2003) Supporting decentralised urban governance: training women municipal councillors in Mumbai, India. Working paper series no 386. Institute of Social Studies Netherland. scholar.google.com/citations?user=6iT1FRIAAAAJ&hl=en. Accessed 30 Jan 2014

Horowitz L (2009) Getting good governance for women: a literature review. Agriculuture and rural development, discussion paper 43 World Bank, pp 39–45. siteresources.worldbank.org/EXTARD/.../GenderandGovernanceLitReview.pdf. Accessed 7 Nov 2012

Hossain M (2012) Influence of social norms and values of rural Bangladesh on women's participation in Union Parishad. Indian J Gend Stud 19(3):394–412. https://doi.org/10.1177/097152151201900303

Iyer L, Mani A, Mishra P, Topalova P (2011) The power of political voice: women's political representation and crime in India. Working Paper11-092, Harvard Business School, pp 2–34. www.hbs.edu/faculty/Publication%20Files/11-092.pdf. Accessed 26 Sept 2012

John ME (2012) Women in power? Gender, caste and the politics of local urban governance. In: Raghunandan TR (ed) Decentralisation and local governments. The Indian experience essays from economic and political weekly. Orient Blackswan, New Delhi, p 349
Joseph V (2011) Capability of elected women representatives on gender issues in grassroot governance. Concept Publishing Company, New Delhi
Khan NI (2014) Gender and local governance: experiences of women representatives in Bangladesh. Ph.D thesis. Department of Political Science, University of Dhaka. Institutional Repository. http://repository.library.du.ac.bd:8080/bitstream/123456789/712/1/Nazneen%20 Islam%20Khan.pdf. Accessed 21 Mar 2017
Khattak SG (1996) Women and local government. Working paper series 24. Sustainable Development Policy Institute, pp 3–16. https://sdpi.org/publications/files/W24-Women%20 and%20Local%20Government.pdf. Accessed 29 Jan 2013
Kodoth P, Mishra US (2011) Gender equality in local governance in Kerala. Econ Polit Wkly XLVI(38):36–43. https://www.epw.in/journal/2011/38/insight/gender-equality-local-governance-kerala.pdf. Accessed on 02 July 2012.
Kulkarni S (2011) Women and decentralised water governance: issues, challenges and the way forward. Econ Polit Wkly XLVI(18):64–72. https://www.epw.in/journal/2011/18/review-womens-studies-review-issues-specials/women-and-decentralised-water
Kumar NA (2006) Do women have efficiency to run the urban administration? In: Singh UB (ed) Empowerment of women in urban administration (experiences and strategies). Serials Publications, New Delhi, pp 105–114
Nair J (2012) All in the family? Gender, caste and politics in an Indian metropolis. In: Banerjee N, Sen S, Dhawan N (eds) Mapping the field (Gender relations in contemporary India) volume 2. Mandira Sen for STREE, Kolkata, pp 464–497
Nanda A (2012) Engaging with democracy: women in panchayats. In: Palanithurai G (ed) Elected women representatives and empowerment. Concept Publishing Company, New Delhi, pp 40–55
Narain K, Mridubhashini M (2006) Dynamics of empowering women in urban development. In: Singh UB (ed) Empowerment of women in urban administration (experiences and strategies). Serial Publications, New Delhi, pp 184–193
Omvedt G (2005) Women in governance in South Asia. Econ Polit Wkly 40(44–45):4746–4752
Palanthurai G (2012) Elected women representatives and empowerment. Concept Publishing Company, New Delhi
Palanthurai G, Uma G, Vanishree J (2009) Networking of elected women representatives at grass-roots. Concept Publishing Company, New Delhi
Panday PK (2008) Representation without participation: quotas for women in Bangladesh. Int Polit Sci Rev 29(4):489–512
Paradza G, Mokwena LG, Richards R (2010) Assessing the role of councillors in service delivery at local government level in South Africa. Rep. No. 125. Centre for Policy Studies, Johanesberg, South Africa. www.cps.org.za. Accessed 28 May 2013
Rizvi SH (2006) Empowering women through reservation in urban local bodies-difference between theory and practice. In: Singh UB (ed) Empowerment of women in urban administration (experiences and strategies). Serials Publications, New Delhi, pp 193–211
Ruet J, Lama-Rewal S (2005) Governing India's metropolises. Routledge, New Delhi
Samanta G, Banerjee R (2013) Rajniti O Prashasan Ebong Nari (Politics, governance and women). Antarmukh, January–March, 15–25
Samanta G, Banerjee R (2016) Bibi-Bahu-Beti' Brigade Periye (Beyond wife, daughter-in-law and daughter Brigade). Arek Rakam 4(16):38–39
Samanta G, Sil P (2013) Negotiating gendered urban space: experiences of women across classes in India. In Marius K, Raiband Y (eds) Genre et Construction de la Gepgraphie. Maison des Sciences de l'Homme d'aquitaine (MSIA), France, pp 209–222
Sathe D, Klasen S, Priebe J, Biniwale M (2013) Can the female Sarpanch deliver? Econ Polit Wkly XLVIII(11):50–57

Sekhson J (2006) Engendering grassroot democracy: research, training and networking for women in local self-governance in India. NWSA J 18(2):101–122. http://www.jstor.org/stable/4317209. Accessed 19 June 2013
Sen A (2000) Development as freedom. Oxford University Press, New Delhi
Shah AS (2015) Gender polarization in Pakistan Parliament. The Express Tribune. https://tribune.com.pk/story/864544/gender-polarisation-in-pakistans-parliament/. Accessed 26 Nov 2013
Sheel R (2003) Women in politics in Japan. Econ Polit Wkly 4097–4101. http://www.epw.in/journal/2003/39/commentary/women-politics-japan.html. Accessed 19 Dec 2012
Sheridan A, McKenzie FH, Leonie S (2011) Making visible the 'space of betweenness': understanding women's limited access to leadership in regional Australia. Gend Place Cult J Fem Geogr 18(6):732–748
Singh SK (2006) Women empowerment in India: emerging leadership in urban U.P. In: Singh UB (ed) Empowerment of women in urban administration (experiences and strategies). Serials Publications, New Delhi, pp 61–64
Srivastava DN (2015) Political status of Indian women: progess since independence. In: Sarkar S, Pandey PK (eds) Women in politics in South Asia. Mangalam Publications, New Delhi, pp 199–213
Thakkar U (2006) Being women councillors [Review of the book Democratisation in progress: women and local politics in urban India by Archana Ghosh and Stephanie Tawa Lama-Rewal]. Econ Polit Wkly 41(8). https://www.epw.in/journal/2006/08/book-reviews/being-women-councillors.html
Upadhyay KK, Bhardwaj P (2006) The gender paradigm in local governance: Indian experience. In: Singh UB (ed) Empowerment of women in urban administration (experiences and strategies). Serials Publications, New Delhi, pp 130–148
Yupanqui AM (2011) Amartya Sen's notion of freedom a conceptual history of a universalist presumption. SIMT29, Master's thesis in developmental studies. Faculty of Social Science, Graduate School, Department of Political Science, Lund University: Spring. lup.lub.lu.se/student-papers/record/1966976/file/1976430.pdf. Accessed 2 June 2016

Chapter 7
Women and Men Councillors: Comparison of Functioning

'Urban governance in turn can only be effective with the involvement of women alongside men. Women and men experience and use the urban environment in different ways and often have different priorities in terms of urban services and infrastructure. These differences can only properly be understood with reference to prevailing constructions of gender'. – UNCHS 2000: 11

7.1 Introduction

The capability of the councillors is analyzed by their freedom and functioning behaviour. In the present research, men councillors (MCs) are also selected as research participants to analyze their perception about their co-workers, that is, the women councillors (WCs), and to compare their freedom and functioning behaviour. This explanation will enable us to answer the question: How far away are WCs from equality in case of freedom-functioning behaviour from MCs? In search of the answer to this question, the present chapter makes a comparison of capability between men and women councillors in the selected cities of West Bengal.

7.2 Case of Darjiling

In Darjiling, WCs are elected from unreserved as well as reserved seats. Except for one, all the MCs are elected in unreserved seats. WCs and MCs both have first-time councillors. MCs have individual reasons for being elected as councillors. For example, MC 25 said that he is in this field to save the non-Gorkhas. MC 12 believes that after retirement, work of the councillor is a very good job as it is related to public services. MC 18 is a CIC councillor, and got elected to represent the demands

R. Banerjee, G. Samanta, *Negotiating Terrain in Local Governance*, Local and Urban Governance, https://doi.org/10.1007/978-3-030-60663-3_7

of the Muslim community. He comes to give voice to the Muslim community and to fulfil their demands. Therefore, there are political as well as communal reasons behind the nomination of the MCs. However, WCs usually have more emotional reasons behind their nomination, such as to secure their Gorkhaland for future generations.

The MCs of Darjiling have less exposure in case of political membership in other organizations than the women councillors. Language differences also create a barrier among the MCs, just like in the case of the WCs. For example, MC 25 (Bengali-speaking MC) said that he has given various kinds of ward development proposals to the municipality, but they were not granted. His ward remains undernourished as a result, and the local people also blame him. He also added that there is oppression in meetings, which is why he does not attend the meetings.

For the development of the wards, the division between CIC councillor and general councillors is also prominent. MC 28 (a CIC councillor) said that he has developed a five-year plan and given it to the municipal authority to generate funds. He has also visited the municipal affairs department in Kolkata for funds. However, MC 25 has a completely different viewpoint regarding the same. He said that there is no significance of the board of councillors, because CIC councillors have the authority to take all the decisions compared to a general councillor. As a result, more development works are undertaken in their wards.

In case of socio-economic freedom, the level of education of the MCs is quite significant. Three of them have completed *Madhyamik*, two have graduated from higher secondary, one has a graduate degree, and one holds a Master of Arts degree.

In case of economic opportunities, five men councillors were engaged in jobs before becoming councillors. One among them is a retired bank officer, MC 28 was in the military, MC 21 is a *Panchayat* clerk, MC 28 is a coach in sports, and MC 9 is a primary teacher. Though MC 28 has an MA degree, he runs his own tailoring shop and takes tuition for economics. The economic opportunity for the MCs is higher than the WCs.

The physical mobility and media exposure of the MCs is also limited like the WCs. It has been observed that due to family responsibility, WCs do not want to pursue their political career. However, among the MCs of Darjiling, the urge for pursuance of political career is very less. The reason is mainly political.

Therefore, in Darjiling, the difference between WCs and MCs is less in terms of political and socio-economic freedom. The reason for this is the comparatively smaller gender gap in society. As a result, women and men face similar positive and negative situations in the political field.

MCs are also disappointed about their salary. For example, MC 26 has said that the salary is less, but not much less, because it was just INR 200 earlier. MC 28 with a lot of disappointment said that,

> 'We are doing work daily at the ward without any travel allowance and medical facilities and our salary is also very low (INR 1500 per month). Roshan Giri (the then General Secretary of Gorkha Jana Mukti Morcha) by showing our works takes the votes and has the salary of INR 40,000 per month'.

A similar observation is found in the study of Ghosh and Lama-Rewal (2005) in Kolkata that the government spends INR 80,000 per month for a single Member of the Legislative Assembly for his or her personal assistant, car, cell phone, etc. Such facilities should be extended to the councillors because they do the maximum work.

In this context, MC 25 mentioned that there is no honorarium prescribed in the West Bengal Municipal Act. According to the act, there are only seven members in the CIC, including the chairman and vice-chairman. However, the number of CIC councillors has doubled here to acquire the positions of power. In case of honorarium, seven councillors get the salary as CIC members (INR 2100 per month), and the rest have the salary of the general councillors even if they are in CIC position. For example, MC 18 and MC 28 are both CIC members. However, MC 18 gets an honorarium of INR 2100 per month, whereas MC 28 gets only INR 1575 per month. To hold the power position, most men councillors are included in the CIC; however, the honorarium of the seven CIC councillors is fixed. It leads to disparity in case of honorarium among the CIC councillors. Therefore, acquisition of power is important for the ruling political party, which is why the number of CIC councillor is higher in case of Darjiling.

7.3 Case of Balurghat

MCs of Balurghat are also experienced members of different political parties. For example, MC 7 and MC 17 are in the Revolutionary Socialist Party (RSP); MC 4 is a member of Communist Party of India (Marxist); and MC 7 was in the Congress earlier, and at present, he is in the All India Trinamool Congress. On the contrary, WCs are from the same political party background, i.e., the RSP.

Though WCs acknowledge the support of their woman chairperson, MCs think the chairperson was autocratic in decision-making. This attitude signifies the patriarchal thinking of the society at large towards women. As they are not habituated to accept the decisions of women in their families, they feel uneasy following the decisions given by the chairperson.

In case of socio-economic opportunities, most of the men councillors in Balurghat are educated up to graduation level. Among them, MC 20 is a businessman and his wife is in service. As his wife is also working, the family is not solely dependent on him economically, and he is relatively free from these responsibilities. MC 7 is also in business, while MC 17 is a retired primary teacher, and lastly, MC 4 is a computer teacher in a school.

The physical mobility of the MCs is higher than that of the WCs. MC 7 has visited Delhi and MC 17 visited Siliguri. In case of confidence to speak in public, the MCs are confident while speaking in public. Being men in a patriarchal society, they do enjoy greater freedom in public spaces. However, the men councillors in Balurghat have to maintain some norms within the private sphere of their family. As for example, the father of MC 20 has developed a rule regarding the entry time to the house at night, which should be within 10 pm to 10.30 pm. MC 7 also maintains a time schedule of coming back to his house in the night to avoid misunderstandings with his wife. These problems are less among the women councillors (WCs) as most of them are either widowed or unmarried. Men councillors are satisfied with the power and respect they have received from the people. Though WCs have a positive vision of the future, the MCs have a more negative vision regarding this. The reason is again political (like the MCs of Darjiling). MCs said that the recent political situations, where there is no opposition party and political grouping within the same political party, reduce their interest in a political career. Besides their political identity, they have membership in social and cultural organizations such as the Red Cross, 'Kochi Club Academy' and 'Natya Mandir'.

The interpersonal relationships among the MCs are good. However, MCs have some negative views about the chairperson. They think that the chairperson gives greater preference to the WCs and other political members of her own party. They also acknowledge that she is very active in the work as the chairperson of the city.

MCs are also very active in doing the work related to the development of basic services such as water supply, drains, roads, electrification, etc., as well as implementation of other developmental works such as establishment of the 4G tower, giving trade license, implementation of *toto rickshaw* service and so on, along with empowerment programmes such as Below Poverty Level cards, *Swarojgar Yojana*, training related to mobile repairing, stitching, small industries work, etc.

Therefore, in Balurghat, MCs are also active like the WCs. However, MCs have complaints against the woman chairperson as she gives more opportunities to WCs. This reaction is due to the inherent deeply rooted patriarchal mindset of the MCs.

7.4 Case of Raniganj

In case of socio-economic freedom in Raniganj, the educational levels of the MCs are quite high. Except for two (MC 10 and MC 15), all of them have a graduate or higher secondary degree. MCs of wards 6 and 19 have graduate degrees.

All MCs have temporary sources of income such as a small business, tuition and temporary work, etc. In this context, MC 10 said:

> 'I am a rickshaw-puller, elected as a councillor as party selected me. However, I have to do my daily job to feed my family. I have no time to do work as a councillor. I am a common man who is struggling daily to feed the family'.

MCs have limited access to media like the WCs have. Government reports and newspapers are also not available in the meeting place, and are only available in the chairman's cabin. Without the permission of the chairman, it cannot be accessed by others.

In case of membership of social organizations, MC19 is associated with youth associations and MC 3 with a sports club. Others do not have attachment either to social organizations or to NGOs. All the MCs acknowledge the support they get from the chairman. In contrast, for the WCs, the chairman acts as an influential person causing identity crisis and discrimination.

In Raniganj, MCs acknowledge the support they get from their party, and they do not face any kind of oppression in the party, unlike WCs who face more opposition from their own concerned party as well as from the dominant party. Thus, within the same party structure, women face greater stress, whereas men get support. This observation clearly indicates the role of patriarchy in play within the society. WCs also have less exposure than the MCs in case of membership of other organizations. MCs are attached with other political organizations such as different *samities* and *sanghas*.

MCs are not excluded in the meeting like the WCs are. Chairman gives opportunities to the MCs to raise their issues in the meeting. However, WCs face party-based opposition in meetings as well as discrimination in case of allotment of funds for the developmental works.

About their honorarium, all of them said that it was INR 750 earlier and, after that, it has been increased to INR 1000. However, according to their perception, it is a very less amount. In this context, MC 19 said that it will be better if it is increased up to INR 2000 per month. MC 3 argued that the financial condition of the municipality is not very good, and as a result, the honorarium is low, because of which they have to spend money from their pocket to do social service.

Therefore, there is huge gap in between MCs and WCs, in case of opportunities and freedom in Raniganj. The WCs face more discrimination and opposition in the municipal system in Raniganj, which was not observed much in case of cities in north Bengal such as Balurghat or Darjiling.

7.5 Case of Hugli-Chinsurah

In Hugli-Chinsurah, the social freedom in terms of education and occupation is good among the MCs. Among them, one has an MA degree, two are graduates, and the rest have studied till higher secondary level. Except for one government employee (MC 22), all are engaged in private occupations, that is, doing business and medical representative's work.

The level of physical mobility of the MCs is also very high. However, in case of WCs, the physical mobility is very low. MC 14 has considerable physical mobility, having travelled within India, and three other MCs are mobile within the state. They are confident while speaking in the public forum as they think that they have family-like relations with the people of their wards. For these reasons, most of them want to pursue a political career to work for the people. Except MC 7, all of them come under the purview of the media during the inauguration of developmental work, during rallies and also during incidents such as floods and so on. They are also engaged in other social organizations such as sports clubs and youth associations.

All of them are also quite aware of government schemes. They have implemented *Rajiv Awas Yojana*, different grants from central and state government (for Below

Poverty Level households, old age, students, unemployed and widowed) and BSUP and undertaken construction of houses under *Gitanjali* project, health schemes, etc. However, the exposure of the MCs in case of membership to other organizations is very low.

The MCs are elected in unreserved seats except MC 10. He is elected to the seat reserved for SC. In case of WCs, they are elected in the seats which are reserved mainly for the women to secure the seat for men in the future. Among the seven interviewed MCs, four have work experience as councillor for two tenures. Among the experienced MCs, two are in chairman-in-council (CIC) position. Therefore, experienced MCs have been given the CIC position. There are experienced women councillors, but they are not in the CIC. Thus, discrimination based on gender in accessing power is very common in Hugli-Chinsurah.

MC 18 acts as a store in-charge, and that is why he comes to the municipality on a daily basis. MC 22 has the responsibility of the department of health. In this context, it is important to mention that according to Kodoth and Mishra (2011), women representatives are placed in less important departments such as health, child care, etc. However, in Hugli-Chinsurah, men councillors are also placed in the department of health.

The political party-wise distribution shows that two are from Communist Party of India (Marxist) (ruling) and four are supporters of All India Trinamool Congress (opposition). There are also MCs who are from the opposition party. However, according to the MCs, there is no oppression in the meetings. Therefore, party-based opposition is more active towards the women councillors (WCs), though MCs are not victimized due to that opposition. In case of WCs, there is no formal oppression in meetings. Because WCs are elected to proxy seats, they have to present themselves in the meeting just for attendance. However, the decisions and implementations are done by their husband or associated party members. Another issue of discrimination among the WCs is that whenever the WC from the opposition party raises any issues in the meeting, nobody listens to her.

Therefore, in Hugli-Chinsurah, MCs do not face discrimination in meetings. Most of them are also in the CIC position. On the other hand, the WCs are elected in the proxy seats; they are discriminated against in their own decision-making, and are excluded from the post of CIC councillor, even though some of them have the requisite work experience.

7.6 Perception of Men Councillors About Women Councillors

Good attitude and perception of the men councillors (MCs) towards the women councillors (WCs) play an indirect role in the freedom of the women councillors in the governance field. The positive outlook of the MCs makes the path of the WCs much easier, enabling them to reach the ultimate capability level in all fields of their work. It also helps the men councillors to do their work with their co-workers (women) without any hesitation.

In **Darjiling**, MCs have a positive attitude about the work of the women councillors. For example, MC 18 thinks that women are more suitable for indoor work, but they face some problems in outdoor works, and that is why they need more support in the outdoor activities. MC 26 thinks that men and women both should have equal opportunities in this field. However, beginners should have the opportunities to learn. MC 28 hesitates to talk to the women in general. This is the main cause of miscommunication with the WCs. He also pointed out the disadvantages of the WCs of Darjiling. One is non-attendance in the municipality daily due to household chores. Second, as most of them are beginners and municipality staff is already experienced, that is why it creates problems. MC 25 narrated both the advantages and disadvantages of the WCs in Darjiling. The advantage is that they are in a better position in the society as women. They enjoy power and socio-economic position. The disadvantages are that most of them are beginners with lack of maturity and lack of knowledge and education. They have limited space in politics, and sometimes the family, husband and political party misuse the power position of the WCs.

In **Balurghat**, MCs have a positive attitude about their women co-workers. However, they have some complaints against the woman chairperson. They feel that the chairperson practices autocracy in case of placing decisions. MC 20 does not respect the WCs elected in reserved seats. MC 20 feels that women in reserved seats are not eligible for the post of the councillor. This kind of negative viewpoint of the MCs reveals their inherent patriarchy as well as a deeply rooted sense of social hierarchy. It is a fact that experienced women party members sometimes do not get tickets for nomination from their party. It is because of reservation that they have the opportunities to be nominated as a councillor. Therefore, in Balurghat, though the encouragement of the chairperson helps the WCs to get more visibility, negative viewpoints of their co-workers (MCs) are a limitation.

In **Raniganj**, though the discrimination and opposition are prominent against the WCs, most of the MCs have positive viewpoints about the WCs. They feel that WCs have different arguments for the development of the wards in board of councillors meetings. They have good decision-making qualities. As co-workers, WCs are helpful and maintain very good relations with the MCs. However, MC 19 said that it is very difficult to speak with the WCs freely. This miscommunication is also a problem between men and women councillors as it was also observed in the case of Darjiling.

In **Hugli-Chinsurah**, MCs have different viewpoints about the WCs. Except two MCs, the rest think that there is no disadvantage of having women as their co-workers; however there are advantages, such as getting help and discussing some issues with them. Two men councillors (MC 3 and MC 7) highlighted the disadvantages of working with the women councillors. According to them, WCs do not come to the meetings regularly, and they do not understand their work properly, and as a result, the speed of their work is very slow. Most of the men councillors feel that WCs have the burden of their house as they are housewives, which is why they are not able to take the workload of a councillor. They also think that WCs face tremendous crisis of time for their work as councillors, and that is why they sometimes perform badly.

7.7 Social and Cultural Context Matters

The level of political freedom of the women and men councillors is more or less same in case of Darjiling. However, the educational status and economic freedom of the men councillors are much higher than that of women councillors. Women councillors are elected because they are motivated by the emotional issue of saving Gorkhaland, but the men councillors are elected because of different communal issues, such as to save the identity of Muslims and non-Gorkhas in the would-be Gorkhaland. In Darjiling, men councillors have a largely positive perception about their women councillors. They think that women are active and responsible.

In Balurghat, all the men councillors enjoy their social, political and economic freedom. They also acknowledge the influence of the woman chairperson in bringing freedom in case of decision-making in meetings. However, one among them has a very negative perception about the women who are elected through quota. In Raniganj, women councillors faced many hurdles from the members of their own party. But in case of men councillors, they enjoy their freedom. In spite of having higher educational level, their socio-economic freedom is less. In case of perception about the women councillors, men councillors are quite positive; however, women councillors are mostly deprived and controlled here. In Hugli-Chinsurah, men councillors enjoy all kinds of freedoms, whereas women councillors are in proxy seats, and are deprived of the opportunities to take decisions. Men councillors also perceive that the household burden is more important for the women than other work.

Therefore, this comparison between WCs and MCs reveals that the differences among them in terms of freedom also depend on space specificities. Different geographical spaces with different cultural set-ups determine the individual perception. Because of the progressive cultural set-up, the perception of men councillors about the women councillors is more positive in the cities of Darjiling and Balurghat situated in the northern parts of West Bengal. However, in the cities of Raniganj and Hugli-Chinsurah, situated in the southern part of West Bengal, the discrimination, opposition and proxy representation of WCs are very high. The negative perception of the MCs signifies their deeply rooted patriarchal mindsets.

After identifying the perception of MCs on the performance and capability of women councillors in this chapter, the next chapter analyzes whether the citizens perceive the capacity of men and women councillors differently or not.

References

Ghosh A, Lama-Rewal ST (2005) Democratization in progress (women and local politics in urban India). Tulika Books, New Delhi

Kodoth P, Mishra US (2011) Gender equality in local governance in Kerala. Econ Pol Wkly XLVI(38):36–43. https://www.epw.in/journal/2011/38/insight/gender-equality-local-governance-kerala.pdf Accessed on 02 July 2012

UNCHS (2000) Policy paper on women in urban governance. Draft paper by UNSCH (Habitat). http://www1.aucegypt.edu/src/engendering/Documents/Home%20Page/women%20and%20urban%20governance.pdf. Accessed on 20 November, 2000

Chapter 8
Citizens' Perceptions and Assessment: Gender Matters

> Women and men experience cities differently due to their different roles and activities, and women's needs are seldom represented in policy or planning, it is essential that these interests are now actively advanced (Beall 1996: 2). UNCHS has selected the following areas for strategic attention: basic services, human rights, economic capacity, transport, violence and security of tenure. Issues that affect women are not static, as the roles of women and men in different societies around the world are constantly shifting, especially recently as a result of economic globalisation. (UNCHS 2000: 3)

8.1 Introduction

Perception of citizens is an important factor which helps in identifying the actual capability of a councillor in the field of governance. It acts as a feedback system to know the merits and demerits of the representatives. The study by Amer (2009) on Naga voters has highlighted the judgment and political awareness of the women voters in Nagaland. The study hypothesizes that women voters are the least informed as well as least interested and aware about the field of politics as well as governance. However, Moamenla Amer (2009) has examined and analyzed the Naga women voters, and shown that women's political participation as voters depends on their educational attainment, the pattern of media consumption and frequency of engagement in media discussions. It is also found that awareness is relatively higher among the married women voters within the age of 46–50 years (Amer 2009: 366).

PRIA's (2008) study on small and medium cities has brought a new perspective on citizens' engagement in local governance. It has highlighted the role of governance in enhancing active engagement among the citizens towards governance. For instance, this includes providing services to poor people and utilizing strategies such as development of pre-election voter awareness through campaigns, media, street plays and development of the monitoring committee to analyze the performance of municipal services as well as citizen's engagement.

R. Banerjee, G. Samanta, *Negotiating Terrain in Local Governance*, Local and Urban Governance, https://doi.org/10.1007/978-3-030-60663-3_8

Another important aspect pertaining to the issue relates to the attitude with which women as political representatives are faced. This has been highlighted by Hossain (2012). The study reveals that women who were in the Union Parishad of Bangladesh faced a situation that jeopardized their characters and reputation. For example, some of them were labelled as having loose character, and they faced teasing as they were participating in the field of governance. They faced several hurdles in doing their work. For example, people did not cooperate with women representatives on issues such as eradicating dowry and early marriage of girls and made it clear that they did not prefer women in the role of decision-makers. Though women are in a position to make decisions, people prefer to see them as observers and listeners (Hossain 2012).

Thus, citizens' perception acts like a feedback system to analyze the real nature of functioning capability of the councillors. In the present research, four cities are selected from four geographical regions of West Bengal. With different socio-economic and cultural characteristics, the socio-cultural environment of the concerned cities also varies. It impacts on the problems of the concerned cities, on the functional spaces of the councillors and on the citizens' perceptions. Therefore, to identify the perception of the citizens about the personality and functioning of the councillors, 100 citizens per city have been interviewed. The city-wise breakup and details of citizens covered in this survey are given in Appendices 8.1 and 8.2. This survey targets the household level, and only those people were interviewed who were willing to participate in the research.

To know about the perception of the common people, first of all the political awareness of the common people has been analyzed, to find out the status of their access to political information. Following Amer (2009: 359–360), by 'political awareness' we mean access to political information, political participation, media exposure, interest in politics and education, etc. Citizens should have some basic knowledge about the political matters, on the basis of which they can express their preferences and views about the performance of elected representatives. Some indicators that can be used to analyze the political awareness of citizens are involvement in politics, participation in political discussions, knowledge about the politics of West Bengal, and their access to public media (radio, newspapers, television, etc.).

8.2 Political Awareness Among the Citizens

Political awareness is primarily a function of acquiring knowledge about politics. It means access to political information, political participation, media exposure, interest in politics and education with political information (Amer 2009: 359). It is necessary that citizens should have a minimum knowledge of the political system to express their preferences regarding their elected representatives. Therefore, to analyze the political awareness of the citizens, the following indicators are used: (a) participation in politics, (b) participation in political discussion, and (c) level of education.

The nature of the participation of the citizens in politics is mainly of two types. One is as a voter, and the other is in the process of voting and associated political activities related to election, such as campaigning for a candidate, attending election meetings, party membership and political discussions. The frequency of the citizens' participation in political discussions can be analyzed by the level of occurrence of the participation, such as frequently, sometimes and never.

The level of media exposure of the citizens is indicated by their practice of reading newspapers and watching the news on television. On the basis of these indicators, the political awareness of the citizens has been analyzed. Except for Darjiling, 80% of the men and women citizens participated in politics mainly as voters. However, in Darjiling 50% men and 27% women citizens have participated in other political activities such as attending election meetings and campaigning for the elected candidates. From the discussion on political behaviour of the citizens, it has been found that the engagement in other political activities (camp, rallies, meetings, etc.) is greater among the male citizens. However, in the cities of north Bengal, like in Darjiling and in Balurghat, women's participation in other political activities is also rising day by day.

The frequency of participation of citizens in political discussions shows that more than 60% men in Balurghat, Raniganj and Hugli-Chinsurah have participated sometimes in the political discussions, whereas in Darjiling 30% men participated frequently in political discussions. Except the women of Darjiling, women in all other cities are less interested in participating in political discussions. In Darjiling more than 50% of women have participated sometimes in political discussions, whereas in the other three cities (Raniganj, Balurghat and Hugli-Chinsurah), more than 70% women never participated in political discussions.

The data thus indicates that women have less interest in politics. The tendency to participate in political discussions is greater among the male citizens than the women. The reason as cited by the research participants during the survey is that men have regular access to forums for political discussions in the public places of their neighbourhood such as in the club, tea shops or road-side benches where they gather regularly. Women have less exposure outside their house to discuss political matters. However, the situation is not absolutely the same in all the cities, due to space specificities. For example, in Darjiling, higher proportions of women are actively engaged in political discussions, and also participate in other political activities, whereas in other cities, it represents a very insignificant percentage.

In case of media exposure, most of the citizens rely on both newspapers and television for information. The habit of reading newspapers is higher among the citizens of Darjiling than in the other cities. The reasons as observed are the inaccessible nature of lands in and around Darjiling, and the very irregular supply of electricity due to landslides and other environmental factors. The citizens of the other cities prefer to use the audio-visual media of television.

However, there are still some men and women who have no direct contact with newspapers and television, even in Darjiling. They depend on other people to get any political information. This proportion is highest among the women citizens of

Raniganj. In Raniganj, the educational level of the participant citizens is low, and they prefer to get the news from other people and remain satisfied with that.

8.3 Citizens' Perception of Women Councillors

The overall perception of citizens about their women councillors (WCs) helps to identify the acceptability of the WCs in the cities. Sathe et al. (2013) have stated that women representatives can perform better to improve service delivery and democratic participation of the citizens in the long run than men representatives. This section on general perception has been developed from the questionnaire survey and the statistical data collected from the survey.

In Darjiling, citizens accept the women as their local leaders. They are satisfied with their leadership qualities. About 70% of the citizens said that their leadership quality is good, and they are accessible to citizens. Therefore, WCs of Darjiling are actively engaged in their work, and this is proved from the citizens' perception. Paradza et al. (2010) also said that the councillor's responsibility is to serve the voice of the people and community needs. In Balurghat, the majority of the citizens are satisfied with their women councillors. They acknowledged their good leadership qualities and good nature and behaviour, as well as their accessibility.

In Raniganj, the citizens are satisfied with the behaviour of women leaders. However, in Hugli-Chinsurah, a larger percentage of citizens identified the women councillors as aggressive, as busy and as leaders without knowledge. As many WCs are elected in proxy seats, the work of the women councillors, especially the outdoor works, is mainly done either by the male members of the family or by party members. The communication with the citizens is much less among the women councillors. Moreover, citizens report that when they go to the councillor's house, they find the WCs busy with their housework. Therefore, proxy nomination creates a vulnerable identity for the women councillors in Hugli-Chinsurah.

Except for Balurghat, in all the cities, there are about 10–20% people who are still disappointed with the leadership quality of the women (Table 8.1). In this context, Paradza et al. (2010) said that councillors are the first point of contact with

Table 8.1 Perception about the leadership quality and nature of behaviour of the WCs

Cities	*Total number of sample*	*Leadership quality (value in percentage)*			*Nature of behaviour (value in percentage)*			
		V. Good	Okay	No capability	Good	Aggressive	Busy	Without knowledge
Darjiling	70	20	70.91	9.09	90.91	3.64	5.45	
Balurghat	56	77.27	20.45	2.27	93.18	6.82		
Raniganj	56	88.75		11.25	96.25			3.75
Hugli-Chinsurah	63	66.67	13.33	20	64.44	28.89	2.22	4.44

Source: Field Survey, 2013–2015

citizens, and their performances really matter to the people. Whenever there is doubt that the councillor is not performing the duties, it may make the citizens angry. The level of anger becomes higher if the councillor is a woman, as the citizens have in ingrained mind-set that women are not capable of doing public work.

The daily ward visits of the WCs are higher in Balurghat and Raniganj. In Darjiling, WCs mainly either visit their ward once a month, or visit the ward to solve specific problems such as if there is any accident or landslides (Table 8.2). In Hugli-Chinsurah, it has been found that the councillors visit the ward on a daily and weekly basis. However, from the discussion in Chap. 5, it has been found that many WCs are in proxy seats, and the men of their families or parties do the work. Therefore, citizens sometimes do not know who the real councillor is, and they accept the concerned men as their councillors. At this point of discussion on the performance of the WCs, we are also in a fix. If we consider the regular visits as an indicator of performance, then it is good. However, these visits are not being done by themselves, rather by their men family members, which is again bad for women's capability enhancement.

The work of women councillors in urban governance has an added advantage as they can take initiatives for women's empowerment programmes. As women, they have in-depth understanding regarding the discrimination against women in the society. It is also one of the agendas for women's reservation. However, it is very shocking to learn that in all the cities, women councillors are least concerned about women's empowerment programmes (Table 8.2). In Balurghat and Raniganj, about 50–60% of the people are quite satisfied with the women's empowerment programmes taken up by their women councillors. From the previous discussion, it is clear that Darjiling and Hugli-Chinsurah have a different setup regarding the nature of freedom of the women councillors. In Darjiling, the numerical representation as well as position of the WCs in administration is higher, whereas in Hugli-Chinsurah, most of the women are elected to proxy seats, and their nature of freedom is obstructed by male-dominated politics. However, the negative responses regarding women's empowerment programmes taken by the women councillors is the same, and more than 70% people said there is no initiative taken by the women councillors. The reason for this, however, is observed to be different in the two cities. In Darjiling, non-availability of funds and less education among the WCs are the main reasons for non-execution of women's empowerment programmes. In Hugli-Chinsurah, though the WCs are educated, the proxy representation is the reason for non-execution of the said programmes.

Lastly, to analyze the capability of the women councillors, a score card regarding the work of the women councillors has been developed. In the score card, there are marks from one to five, where the higher value represents higher capability. The marks are given by the citizens of their concerned wards. The result shows that in all the cities, the percentage is high towards the marks 'three', that is, women councillors have moderate capability to satisfy the needs of the common people. There are a percentage of people who gave five marks to their women councillors. It signifies that they have faith that women can do better (Table 8.3).

Table 8.2 Perception about the functioning of the WCs

Cities/total number of sample	*Accessible citizens (value in percentage)*		*Visit to ward (value in percentage)*					*Have ability to solve the problems (value in percentage)*		*Concerned about women's empowerment (value in percentage)*	
Percentage distribution	Yes	No	Daily	Weekly	Monthly	Never	Based on problems/ sometimes	Yes	No	Yes	No
Darjiling (70)	70.9	29.09	3.64	12.73	29.09	19.05	16	40	60	25.45	74.54
Balurghat (56)	100		43.2	43.18	13.63			77.27	22.7	50	50
Raniganj (56)	100		57.5	42.5				83.75	16.2	63.75	36.25
Hugli-Chinsurah (63)	93.33	6.66	26.7	48.89	6.66	20		73.33	26.67	28.89	71.11

Source: Field Survey, 2013–2015

Table 8.3 Perception about the capability of the WCs

Cities	*Total number of sample*	*Given marks to the women councillor by common people of their ward* Given marks and perception of the common people in percentage				
		1	2	3	4	5
Darjiling	70	16.36	10.91	36.36	18.18	20
Balurghat	56		2.27	61.36	31.82	4.54
Raniganj	56	2.5	7.5	50	32.5	7.5
Hugli-Chinsurah	63	11.11	11.11	31.11	35.56	11.11

Source: Field Survey, 2013–2015

8.4 Digging Deeper into the Perceptions

This section analyzes the perception of citizens on the capability of women councillors on the basis of individual cases. In this section, the analysis is made on the basis of cities differently, as the perception of citizens varies widely from city to city. The perception itself emerges out of the long-standing social norms with which the citizens have grown up. This part is developed from the field work using qualitative methods focused on the narratives of the people, that is, the qualitatively expressed perception of the citizens on their individual women councillors. The analysis is done on the basis of individual councillors, as assessed by the citizens of their respective wards.

8.4.1 Case of Darjiling

In Darjiling, 15 wards have women councillors (WC). In ward number 1, people are disappointed with the councillor. They said that she did not do anything. There is no development of basic services such as water supply, construction, and repairing of roads. Street lights are there, but they are not in working condition. Municipality workers do not come to the ward for garbage cleaning. They reported to the ward councillor, but she did not take any initiative. The perception is same among the common people in ward 2. They said that she is a political person, and only comes during the time of election. Except for giving the residential certificates, she does not do anything. The area is landslide-prone, but she never facilitates them during such hazards. In ward 3, people do not know the councillor, but they do know that she is a Tibetan woman. Garbage is a serious problem of that ward. The reason for such non-implementation as noted during the interviews with the councillors is that these wards are located quite far from the Municipal office. WCs do not go to the office daily which is already discussed in Chaps. 4 and 5. Therefore, they do not get the information about the ward development schemes, and do not have the opportunity to place their demands for the development of the wards in the Municipality.

Distance is the major factor discriminating the development of the wards which are located in the peripheral areas, and far from the city centre.

The WC of ward 5 tries to develop the condition of her ward. People have faith on her. If they complain, the WC tries to solve that problem. She has developed garbage cleaning facilities, and wants to develop a public toilet, but there is no proper land for that. In ward 7, people are also disappointed with the ward councillor. One participant clarified his argument about the unequal ward development in his statement as follows:

> She arranges the ward committee meeting just outside her house because of her restricted mobility. The ward is big and young people usually remain outside home for work. As a result of this, elderly people have to attend the meeting which becomes hard for the old men and women to walk through the steep roads of hill to attend the ward committee meeting. Because of this, the development does not take place uniformly, only the people who attend the meeting from her neighbourhood get the benefit of the ward development program.

People are satisfied with the work of WC 10. She has undertaken road development and repairs in a 1-year interval. She established a public toilet and increased services to clean the garbage. For women's empowerment programmes, she has spread awareness among the women regarding the Integrated Child Development Services (ICDS) programme. She remains active in solving the problems of the people of her ward. The people of ward 13 said that their ward councillor is uneducated, and that is why she is unable to write an application regarding the development of her ward. As a result, the ward remains underdeveloped. Therefore, less education among the WCs is one reason for less development in their ward. Due to less education, they cannot even write an application to the higher authority to place their demands to solve the problems. WC 14 tries to develop her ward. She takes initiatives regarding grants for the poor, and old men and women. She has improved the road conditions and garbage cleaning facilities. However, the size of the ward is large, and it is very hard for her to ensure inclusive development of her ward.

People are highly satisfied with the work of WC17. In their words:

> She does not know us but she has come if we message her regarding any problem. She never harassed us and does our work in the Municipality as early as possible. She has developed the electricity condition and constructed the public tap. Municipality workers come daily to clean the road and toilet.

WC 20 has also taken many initiatives, and the work is in running condition. WC 27 has also developed footpaths, and public toilets, and established houses under the scheme of *Indira Awas Yojana* as narrated by the citizens.

8.4.2 Case of Balurghat

In Balurghat, eight WCs are elected from different wards. Among them, WC 1 is the Chairperson of the Municipality. As a Chairperson and as a political party member, she is famous in Balurghat as well as in her ward. Citizens are satisfied with her work. She has developed the condition of water supply, toilet facilities, and garbage cleaning services. However, road construction work is in process. Citizens have acknowledged that she is accessible to the citizens and has the ability to solve their problems.

Citizens of ward 3 said that their ward councillor behaves well. However she depends on political members. The ward is large in terms of area. She has developed the basic services, though slum areas remain undernourished, and during floods the area goes under water. The garbage cleaning facilities are also very poor. WC 6 visits the ward on a daily basis, and comes to the spot in case of any problem. However, there are a few citizens who are still gender biased and said:

> She is a good woman but she has no leadership qualities like a man. Women come to politics for their own interest not for doing the work for the common people.

This kind of thought is also an obstacle for women councillors to make their identity among the citizens of their wards.

Common people have appreciated the good behaviour of WC 9. She gives time to the people of her ward at home. She has undertaken many developmental works. During the rainy season, the roads remained under water, which is a major problem of the ward. She has provided *tripal* (tarpaulin, a piece of hard cloth) to the people of her ward to make roofs during the rainy season.

Therefore, in other wards of Balurghat, though water supply is a major problem, people are satisfied with the work of the councillors. People believe that women councillors always engage themselves to solve their problems.

In case of women's empowerment programmes, WCs have initiated self-help groups and awareness generation among women. In ward 1, WC has taken an initiative related to encouraging enrolment of girl children in school. In this context, it is important to mention that women councillors are locally important people, and they are more approachable to women voters, as noted also by Thakkar (2006).

8.4.3 Case of Raniganj

In Raniganj, women councillors are elected from eight wards. The people of ward 1, 2, 5 and 9 are satisfied with the work done by their councillors. Most of them have said that there is no problem in their ward. However, few of the people in ward 1 said that there is a problem in drainage, and there is no school in their ward. For this reason, students have to go long distances for their education. In ward 2, one of them said that, 'I am a Below Poverty Level candidate, but she does not provide house for me'. One person from ward 8 said:

> The ward councillor behaves well, but she has no capability in doing work of the councillors. There is no proper planning in our ward, no proper roads, and there is no playground for children.

Therefore, the citizens feel helpless regarding the work of the WC of their ward. Paradza et al. (2010) stated that the anger of citizens may be fuelled if councillors are not performing the work properly. However, people of ward 9 said that though the WC has taken initiatives, these are not implemented. According to the ward people, WC 12 is controlled by the concerned party members. For this reason, she has no capability for doing the work of the councillor.

As discussed earlier in Chaps. 5 and 6, the women councillors of Raniganj face oppression, discrimination as well as opposition, and those are their main causes of identity crisis. Same is observed in the present chapter from the perception of the citizens. The citizens complain about the non-implementation of programs, and the control of the party on the women councillors.

8.4.4 *Case of Hugli-Chinsurah*

Most of the citizens in Hugli-Chinsurah have the mind-set that women have household work, and that is why men are better for the work of the local councillor. This mind-set of the people gets reaffirmed because of the proxy representation of women for the post of councillors. The people do not notice the women councillors directly working for them, as their work is mainly performed by the men of their family or by the concerned political party members. People have many complaints against their women councillors. One citizen said:

> She never visits the wards, she does not do anything, she always is engaged in her housework, she has no time for us; she only puts the signature on paper but does not do any work.

8.5 Overall Findings

The general perception of the citizens about their women councillors has been analyzed quantitatively. The citizens' perception has been analyzed on the basis of their responses about the leadership quality, behaviour, and functioning of the women councillors. From this, it has found that the overall responses of the citizens regarding the work of the women councillors in the cities of Darjiling and Balurghat are good. In the cities of Raniganj and Hugli-Chinsurah, it is moderate. This perception is analyzed on the basis of the quantitative majority of the citizens' responses.

However, while analyzing the detailed perception on the basis of in-depth interviews with the citizens, it has been found that the responses may vary according to individual women councillors in a particular city. From this, it is proved that quantitative analysis can give an overall generalized statement of the facts, but the qualitative explanation is more diversified to clarify the truth behind generalizations.

8.5.1 Citizens' Perceptions of Men Councillors

In the patriarchal society men are highly accepted in every sphere of life. In the political system, how far they are accepted is a matter of analysis. The leadership quality and nature of behaviour of the men councillors is examined on the basis of the perception of citizens. The level of acceptance of the men councillors by the citizens is higher in the cities of south Bengal (Raniganj and Hugli-Chinsurah) than that of Darjiling and Balurghat (Appendix 8.3). From the findings, two hypotheses can be drawn. One is that citizens believe in the patriarchal norms that men are best in whatever they do, and the other is that men councillors have actually done their work very conscientiously.

In case of nature of behaviour, more than 9% of the citizens said that the men councillors are always busy. In Raniganj, 2% citizens confessed that some men councillors are aggressive. Therefore, citizens feel that men councillors are less polite in their behaviour. On the question of accessibility of the councillors, the citizens also added that men councillors usually visit their ward on a weekly basis (Appendix 8.4). In Raniganj and Hugli-Chinsurah, all the citizens think that men councillors have the ability to solve their problems. In Darjiling and Balurghat, a smaller proportion of citizens believe that men councillors can solve their problems. In case of women's empowerment programmes, 97% citizens of Hugli-Chinsurah think that men councillors are aware about the different kinds of women's empowerment programmes, whereas in Raniganj, only 31% citizens have the perception that men councillors are aware of women's empowerment programmes. In Darjiling and Balurghat, these percentages are 66 and 44, respectively (Appendix 8.5).

The perception of citizens about their men councillors in the cities of West Bengal has been discussed. The result indicates that people are more satisfied with the men councillors in comparison with the women councillors in some contexts, while in some other contexts, it is just the opposite. Based on the opinion of citizens, the comparative assessment of the men and women councillors is interpreted in the next section.

8.5.2 Comparison of the Perception of Capabilities of Women and Men Councillors

The opinion of citizens towards their men and women councillors acts as a feedback system. In this context, the study of changing voting pattern of rural West Bengal by Bardhan et al. (2014) shows that there is higher level of dissatisfaction of voters towards their local leaders because of the lack of involvement of them in citizen's services. In the present research, from the comparative assessment of the men and women councillors on the basis of citizens' perception in the selected cities of West Bengal, it is found that women councillors are moderately accepted in all the cities except Balurghat. In Balurghat, people acknowledge the work of their women councillors fully. The acceptability of the men councillors depicts a contrasting picture, as in north Bengal men councillors are equally accepted just like the women councillors, based on their capability to do work for the people. However, in the cities of south Bengal, the acceptance level of the men councillors is of a much higher level. The reason is that local people cannot notice the women working for them as local representatives because of inherent patriarchy, opposition and discrimination based on political party and proxy representation of women. People think that women are busy in their household work and they have no time to work for the society; therefore, men are better as they can give more time to their work.

The most prominent fact is that except in Balurghat and Raniganj, women councillors are not aware of women's empowerment programmes, whereas men councillors are quite aware of it in all cities. In case of nature of behaviour of the women and men councillors, citizens perceived the behaviour of the men councillors as aggressive, and they always represent themselves as busy men. On the other hand, women councillors take the concerned issues emotionally, for instance, by protesting against rape and giving aid during floods and other natural calamities. Their behaviour towards the citizens is always rated as good.

Therefore, it is clear that though the inherent patriarchal mind-set of the local people as well as the system tends to make the women councillors invisible, but their behaviour affects the citizens positively and shows the concern of the women councillors towards them. The next chapter summarizes the entire research; identifies the overall problems of the women in urban governance; discusses the complexity of the problems; and concludes with recommendations of some changes, which can help women function better as councillors and in running the cities.

Appendices

Appendix 8.1: Brief Profile of the Citizens Interviewed

		Brief profile of the citizens						
		Percentage of citizens under the age group of			*Educational attainment of the citizens (%)*			
Cities	*Categories*	*Less than 30 years*	*30 to 50 years*	*50 years and more*	*Illiterate*	*Non-matric*	*Under graduate*	*Graduate*
Darjiling	Men	15	50	35	–	23.5	56	20.5
	Women	12	64	24	4	36	44	16
Balurghat	Men	17	40	43	2	6	52	40
	Women	5	52	43	5	33	19	43
Raniganj	Men	24	41	35	14	55	16	15
	Women	36	52	12	24	48	22	6

Cities	*Categories*	*Brief profile of the citizens*						
		Percentage of citizens under the age group of			*Educational attainment of the citizens (%)*			
		Less than 30 years	*30 to 50 years*	*50 years and more*	*Illiterate*	*Non-matric*	*Undergraduate*	*Graduate*
Hugli-Chinsurah	Men	15	49	36	11.5	51	18	20
	Women	33	57	10	10	57	24	10

Source: Field Survey, 2013–2015

Appendix 8.2: Occupational Status of the Citizens Interviewed

Occupational status	*Darjiling*		*Balurghat*		*Raniganj*		*Hugli-Chinsurah*	
Citizens	*Men (%)*	*Women (%)*	*Men (%)*	*Women (%)*	*Men (%)*	*Women (%)*	*Men (%)*	*Women (%)*
Service	26.5	16	40	24	10	–	17	10
Private jobs	20.5	–	2	–	3	–	–	–
Cultivator	–	–	–	–	1	–	–	–
Non-earner	–	–	6	–	4	4	3	5
Housewife	–	52	–	62	–	34		66
Self-employed	17.6	16	17	9	22	3	18	
Business	35.3	16	26	5	28	3	31	
Others	–	–	9	–	32	6	31	19

Source: Field Survey, 2013–2015

Appendix 8.3: Citizens Perception About the Leadership Quality and Behaviour of the Men Councillors

Cities	*Total sample*	*Leadership quality (%)*			*Nature of behaviour (%)*			
		V. Good	*Okay*	*No capability*	*Good*	*Aggressive*	*Busy*	*Without knowledge*
Darjiling	30	38.09	57.14	4.76	90.48		9.52	
Balurghat	44	79.41	20.59		100			
Raniganj	44	100			97.37	2.63		
Hugli-Chinsurah	37	100			100			

Source: Field Survey, 2013–2015

Appendix 8.4: Citizens Perception About the Work of the Men Councillors

Cities	*Accessible citizens*		*Visit to ward*				
Percentage distribution	Yes	No	Daily	Weekly	Monthly	Never	Based on problems/ sometimes
Darjiling	90.48	9.52	14.3	33.33	4.76	25.45	28.57
Balurghat	100		26.5	50	23.53		
Raniganj	100		31.5	68.42			
Hugli-Chinsurah	100		2.86	97.14			

Source: Field Survey, 2013–2015

Appendix 8.5: Perception of the Citizens About the Capability of the Men Councillors

	Ability to solve problems		*Given marks to councillor marks*					*Concerned about women empowerment*	
Cities	Yes	No	1	2	3	4	5	Yes	No
Darjiling	52.38	47.6	23.81	9.52	23.81	19.05	23.81	61.9	38.09
Balurghat	88.23	11.76			58.82	38.23	2.94	44.12	55.88
Raniganj	100		5.26	2.63	65.79	23.68	2.63	31.58	68.42
Hugli-Chinsurah	100				25.71	51.43	22.86	97.14	2.86

Source: Field Survey, 2013–2015

References

Amer M (2009) Political awareness and its implications on participatory behaviour: a study of Naga women voters in Nagaland. Indian J Gend Stud 16(3):359–374. https://doi.org/10.1177/097152150901600303

Bardhan P, Mitra S, Mookherjee D, Nath A (2014) Changing voting patterns in Rural West Bengal. Econ Polit Wkly:54–62. https://www.epw.in/journal/2014/11/special-articles/changing-voting-patterns-rural-west-bengal.pdf. Accessed on 1 Apr 2014

Beall J (1996) Urban governance: why gender matters? https://www.gdrc.org/u-gov/doc-whygendermatters.html. Accessed on 30 Jan 2014

Hossain M (2012) Influence of social norms and values of rural Bangladesh on women's participation in Union Parishad. Indian J Gend Stud 19(3):394–412. https://doi.org/10.1177/097152151201900303

Paradza G, Mokwena LG, Richards R (2010) Assessing the role of councillors in service delivery at local government level in South Africa (Rep. No. 125). Centre for Policy Studies, Johannesburg, South Africa. www.cps.org.za. Accessed on 28 May 2013

Sathe D, Klasen S, Priebe J, Biniwale M (2013) Can the female Sarpanch deliver? Econ Polit Wkly XLVIII(11):50–57
Society for Participatory Research in Asia (PRIA) (2008) Occasional paper series 9 March, citizen engagement in urban governance lessons from small and medium towns in India, New Delhi. Accessed from https://dspace.library.uvic.ca/bitstream/handle/1828/6531/PRIA_CitizenEngagementUrbanGovernance_2009.pdf?sequence=1&isAllowed=y
Thakkar U (2006) Being women councillors [Review of the book Democratisation in progress: women and local politics in Urban India by Archana Ghosh and Stephanie Tawa Lama-Rewal]. Econ Polit Wkly 41:8. https://www.epw.in/journal/2006/08/book-reviews/being-women-councillors.html
UNCHS (2000) Policy paper on women in urban governance. Draft paper by UNSCH (Habitat). http://www1.aucegypt.edu/src/engendering/Documents/Home%20Page/women%20and%20urban%20governance.pdf. Accessed on 20 Nov 2000

Chapter 9
Gender and Intersectionality: Complex World of Women Councillors

> While the prevalent discourse under the influence of liberal political theorists advocated group-based representation in the form of quotas for 'women', as a single homogenous category, more recent theorization of gender in the context of urban governance visualizes 'women' not as 'neutral de-contextualized citizens'. There is a growing trend towards recognition of multiple social structures (such as gender, caste, class, age, disability, etc.) which often intersect to condition the form and extent of privilege and oppression. In a specific urban context, therefore, gender mainstreaming must accommodate diversities within the group of women and their variegated experiences of city life. (Sethi and Mittal 2015: 1070)

9.1 Women as Councillors

Women's representation at a higher level of government in India is still very insignificant in number (14.58% parliamentarians in the 17th Lok Sabha formed in 2019). However, there are women representatives at higher level of the government, who are influential, and have proved their abilities to become popular leaders. This observation proves that given the chance, women can be competent policymakers, and can run good governments as well. However, the quality of women's representation cannot be ensured by quantitative participation alone. Although there is still no reservation for women at the state and central level of government in India, 33% reservation for women in the local-level governments has been ensured by law around three decades ago. In this context, the present research starts with the argument that without understanding the nature of complexity faced by the women in their daily lives as homemakers and as local representatives, reservation strategies alone are not going to change the sociocultural lives of women in a big way. To justify this argument, empirical research has been done on the changing sociocultural lives of the women representatives, besides their participation in politics and local governance in the selected cities of West Bengal. This section presents a summary

R. Banerjee, G. Samanta, *Negotiating Terrain in Local Governance*, Local and Urban Governance, https://doi.org/10.1007/978-3-030-60663-3_9

of the main findings, which have been explained in detail in the foregoing chapters, while also tracing the overall trends captured by the research.

Vast literature exists on women's participation in local government, with a special focus on the obstacles they face, women's reservation issue, proxy representation of women and citizen's perception on women. A brief outline of the review of literature reveals insights and differential propositions on the themes of women's participation, associated obstacles, and the roles they perform in local governance. Moreover, it also frames the discussion on how far women's reservation is a necessary step in India, besides a comparison with other countries. The citizens' perceptions about women representatives help us in verifying the capability of the women councillors (WCs) as well as their acceptability at the grass-root level. By analyzing the methodologies and findings used in those literatures, the present study developed around the central research question: How are women's social and political spaces changing due to their political participation in the governance of small cities of West Bengal? To address this central research question, a number of objectives and another set of subsidiary research questions were developed. This research focuses on women in urban local governance, systematically analyzes how women negotiate with the terrain of local governance, and explores the barriers and emancipation in the functioning of women. The overarching importance has been given to the local context, which are small towns, where the barriers for women in local governance are more than in big cities.

In exploring the changes in the sociocultural lives and in the nature of political participation of the women councillors, the research starts with understanding the history and consequently tries to look into the historical backgrounds of women's empowerment in each of the selected cities. The analysis of the city-wise women's history shows that out of the four cities, Darjiling and Balurghat, located in northern part of Bengal, have a long history of women's movements and consequent emancipation of women in the society. From the field observation, it has been found that the condition and capability of women councillors in these two cities are far better in comparison to the other two cities of southern Bengal. Thus, it is proved in the process of research that history is important in understanding the present. As a social process, women's emancipation is closely linked with the present status of women in the cities and especially in politics. The present research makes an argument that the positive influence of the history of women's involvement in the public domain helps to increase the emancipation of women councillors and nominate more women even in the unreserved seats.

In understanding the impact of political participation of the women councillors on their sociocultural life, the concepts of having, loving and being developed by Erik Allardt (1993) have been used. From this analysis, the research explores the following changes in the lives of the women councillors. Women's participation in local government increased their work stress and decreased their leisure time to a significant proportion. It also increased the hurdles of their journey, especially in visiting the municipal office and in some cases their own wards as well. Moreover, their normal time-use pattern has changed with their participation as a councillor. They have to manage their time in order to work both in the private and public sphere. In this context, it is important to mention that other women in the family usually help in household work of women councillors. If there is no other woman in the family to help them at home, women face a double burden of work. The nature of the burden depends on their employment status and marital status. Women councillors of all the cities spend on an average 35–40 h in a week for the work of the local councillor. In spite of spending so much time, they do not get any honour, award or prize. The amount of the honorarium is also very low. Therefore, the research makes an argument that lack of financial opportunities in terms of income and the enormous stress linked to management of time between household responsibilities and councillor's work are the major causes of discouragement for women to continue their position in this field. Despite the reservation for women in urban governance over almost three decades, such discouragement does not offer a bright vision for the future of women's meaningful participation in governance.

The research identifies the emancipation status of the women councillors in local governance through the capability approach given by Amartya Sen (2000). The capability of the women councillors depends on the opportunities that help them to get freedom to attain valuable functioning in the political field. By analyzing the freedom and functioning of the women councillors in the cities, it has been observed that the influence of political parties is very strong in the field. Political parties nominate the women for the elected seats (both reserved and unreserved). It has been found in the field survey that the orthodox nature of views of the political parties leads to nomination of only those women who belong to less educated, reserved categories and who have no prior engagement with the party before becoming a councillor. Moreover, councillors' actions are also framed by both the party discipline and electoral calculations. Women councillors have faced discrimination and inherent control by the influential male members of the political parties. They cannot take decisions on their own as it is highly controlled by the party, and if they try to take independent decisions, they face opposition from them.

The associated factors of the influence of the political party are political engagement and political family background. Political engagement of the women councillors from their student lives helps to bring confidence among them. Moreover, it increases their visibility as leaders even before becoming councillor and helps them develop a separate identity apart from councillor. Political family background acts as a positive as well as negative factor for women. In one way, it provides a support base and political knowledge to the newcomers. In contrast, it also encourages the control and interventions of the men of the family over the women councillors. This is also one reason for proxy representation of the women councillors. In some cases, the women councillors are nominated because of reservation, and to secure the seat for the men of their family for the next election, which is called *parivartantra* in Indian politics. From these observations, the research argues that such political complexity negatively influences the decision-making abilities of the women councillors in the meetings and in other public forums.

Moreover, because of opposition and discrimination, their given decisions remain inaudible to the other members and officials in the meeting. The effects of *parivartantra* increase the lack of interest of the women councillors in this field. Women councillors have learnt and received their capability as councillors for their tenure, but they know that they will not be nominated when the seat becomes unreserved. In some cases, women councillors have the mindset that the men of their family and of the political party help them by doing the work of the councillor to reduce their burden. This patriarchal thinking has bound them to the limited space of their house and made them morally inactive, even though they are in the position of councillors. It also decreases the visibility of the women councillors among the citizens, and increases the popularity of the concerned male members for their future career against the same post. Ultimately, it affects the political freedom of the women councillors in the cities.

Such political freedom can be improved by enhancing social opportunities of the women councillors. Social opportunities improve the quality of social interaction. Besides education, physical mobility, training facilities and support from family help the women councillors to gain the confidence to speak in public forums and in front of media. They get opportunities for physical mobility outside their house due to their participation in the political field. This is also a step towards women's empowerment. They become socially important through their ward visits and attending meetings and creating contact with the common people of their wards. Training facilities also help them to know their duties. Therefore, training is necessary during the entry phase of work of the women councillors, as it helps them to develop leadership and communication skills to enable them to function effectively. However, in reality, the effort to arrange trainings for the women councillors is very limited among the officials of the municipal governments. If there are some arrangements for training, most of the women councillors do not attend because of family responsibilities as well as distance of the training venue. There is no compulsion to attend the training in their job rules. From these field observations, the research argues that legal compulsion for training is necessary for the women councillors to ensure their capability.

Family inspiration is a precondition for women councillors to join this field. However, household responsibilities become a major cause for the women not pursuing a career in this field. In this context, the research argues that women councillors need a supportive family and someone in the family who would be willing to take over the responsibilities of the family and children. Besides the support from family, they also need support from their co-workers, including the chairman, other councillors and municipal officials. For this, the research proposes gender-sensitive training for the co-workers of the women councillors. Above all, proper communication and networking among the women councillors help to form a strong group against the traditional male-dominated culture of the offices. Lack of networking among the women representatives makes them more vulnerable to the hidden patriarchal structures within the local government. It is because they have no common voice about their demands that their proposal for development remains unimplemented in most cases.

The different kinds of freedom of women councillors have been explained in detail in Chap. 5. If the women have access to all these political and socio-economic freedoms, they can perform better both as chairman-in-council (CIC) councillors and as general councillors. CIC councillors hold responsibilities of different departments, such as water, land, basic services, poverty eradication, etc. Beside their own wards, they have to take care of the departmental problems as well. They are also in a higher position of the political hierarchy within the local government. The general councillors are responsible for their own ward development. They can take decisions regarding different works, such as providing basic services, developmental works for the ward and income-generating options for the low-income-group people. In case of women's empowerment programmes, it is important to mention that women councillors can understand the women's problem better. They can argue for women's empowerment programmes, can develop health care and can protest against the social problems such as rape and harassment. To know more about all these problems, they visit the wards regularly and attend meetings in the municipal office to solve those problems.

The functioning of the women councillors can also be analyzed through the perception of the citizens as well as of the men councillors (MCs). Perception is the ability to see, hear or become aware of something through the senses to develop an opinion or belief. In the present research, the perception of the men councillors and the citizens about their women councillors has been analyzed. Good attitude and perception of the men councillors (MCs) towards the women councillors (WCs) play an indirect role in the freedom of the women councillors. The positive mindset of the men councillors makes the path of the women councillors much easier, enabling them to reach the ultimate level of capability in all fields of their work. In the cities under study, men councillors have a positive attitude towards their co-workers. However, they believe that household work is more important for the women than other work. It represents the inherent patriarchal mindset among the men councillors of the cities. For this reason, the research already proposes gender-sensitive training for the men councillors as well.

Citizens' perception acts like a feedback system to analyze the real nature of functioning capability of the councillors. Chapter 8 analyzed the patterns of perception of the citizens with respect to the behaviour, capacity, and skill of handling problems and leadership quality of women councillors in running the local government, and the comparison of women on all those counts with the men. The major general observation on women as reported by citizens is that women are good in behaviour as they listen patiently to the complaints raised by the citizens and try to solve their problems as fast as possible. However, citizens do not see women much in the field as there is high incidence of proxy, and the men of the family often do the hard work on the ground. Thus, women councillors' works remain invisible to the citizens. In judging women councillors, citizens are often biased by the standards of the men. That is why women councillors have to prove themselves as competent in terms of the prevailing model, which is the men's model. If we exclude the endemic patriarchal mindset of the citizens, women are actually more easily accepted by the citizens as the people's representative.

However, the general perceptions of the citizens about the women councillors vary widely across cities. Darjiling, being located in the hilly region, faces high water scarcity and landslides as major problems of the citizens. In Balurghat, citizens raised the issues of drinking water quality and floods. In Raniganj and Hugli-Chinsurah, the major demands of the citizens are infrastructural development.

Therefore, the councillors are also engaged to fulfil these demands with their efforts. The overall response of the citizens regarding the work of the women councillors in the cities of Darjiling and Balurghat is good. In the cities of Raniganj and Hugli-Chinsurah, men are rated higher by the citizens as there is huge proxy practiced in these cities of south Bengal. This space specificity is also prevalent in case of freedom and functioning behaviour of the women councillors in the cities.

Therefore, the present research concludes that politics is the source of power and position, and patriarchy is more prevalent there. That is why, even though there is reservation of seats for women at the local government level, the Women's Reservation Bill or the Constitution (108 Amendment) Bill, 2008, which proposed to reserve one-third of all seats in the lower house of parliament and in state assemblies, has never been passed. In local government, reservation helps to increase quantitative representation of women. However, there are many difficulties that women face which have been explained thoroughly in the entire study so far. Therefore, the issue should proceed beyond the number (quantitative representation of women) alone. It is time to focus on the quality of representation of women councillors, besides securing their quantity through reservation. There is also another question to consider—how can the qualitative representation of the women councillors be ensured? In this context, it is important to mention that empowerment of women is not possible if they do not understand their abilities on their own. Therefore, to empower themselves as councillors, it is necessary to provide them with knowledge about the power and responsibilities of the post of the councillors. This can be made possible through sharing of information by compulsory training and other awareness programmes. Even without any formal support for the training and other facilitating programmes from the government, the research has observed that there is a slow process of empowerment continuing among women as a result of their participation in governance. Through participation, women councillors have come in contact with the public sphere outside their homes, and they have developed their own identity in the society and among the citizens. This can be encouraged by giving more economic freedom and moral support to the women councillors. This will secure their freedom and give a good vision of the future among them. To reduce the effects of *parivartantra*, women's work should be more visible to citizens. It will also increase the acceptability of the women councillors to the citizens.

9.2 Intersectionality Within Gender

In doing the field work for the book, we have identified and addressed aspects like education, employment, and political party-wise variation in the capability of women councillors in terms of freedom and functioning across the cities. We have also noted that there is a lot of 'intersectionality' among the women councillors themselves at the individual level. Intersectionality is a term coined by American civil rights advocate Kimberlé Williams Crenshaw to describe overlapping or *intersecting* social identities and related systems of oppression, domination or discrimination (Crenshaw 1991). Intersectionality is the idea that multiple identities intersect to create a whole that is different from the component identities (Intersectionality n.d.; Intersectionality Theory n.d.). The identities that can intersect include gender, race, social class, ethnicity, nationality, religion, age, psychological, and physical ability and other forms of identity (Intersectionality n.d.; Intersectionality Theory n.d.). It provides an insight into how multiple systems of oppression are interrelated and interactive. Intersectionality is not a static field; rather, it is dynamic and constantly developing as a response to formations of complex social inequalities.

The intersectionality framework has helped us to understand how the problems and prospects of women councillors are rooted in their unequal social status, which is again linked to a complex of overarching axes of social determinants such as ethnicity, religion, caste, level of education, age, marital status, and so on. The observed inequality among women councillors in terms of freedom and functioning are rooted in differential axes of factors other than gender only. And there, intersectionality becomes an important factor in making the 'women councillor' a fluid category rather than fixing them into one box opposite to the men councillors.

The diverse social, economic and cultural norms shape women's lives and functioning in ways, which cannot be understood with a simplified lens; rather, we need deeper understanding with appropriate methods to unfold the complex situations within which women councillors are compelled to work. The emancipation of women councillors is a long process, and women are already trying hard to get access to power whenever they get a minimum opportunity to do so. There are some exceptions, but that is not the rule. A little bit of solidarity can help them to fight the battle, which was dominated by men for a very long time. After conducting extensive field work and writing this book, we feel that the term 'empowerment' itself is problematic. Power is not something to give; it is something to grab, and women councillors are fighting for that power within the limited facilitating social, economic, and cultural conditions.

References

Allardt, E. (1993). Having, loving, being: an alternative to the Swedish model of welfare research. In M. Nussbaum and A. Sen (Eds), The quality of life (pp. 88–94). New Delhi: Oxford University Press. https://www.oxfordscholarship.com/view/10.1093/0198287976.001.0001/acprof-9780198287971-chapter-8. Accessed on 10 May 2020

Crenshaw K (1991) Mapping the margins: intersectionality, identity politics and violence against women of color. Standford Law Review 43:1241–1299. https://www.racialequitytools.org/resourcefiles/mapping-margins.pdf. Accessed on 12 June 2017

Intersectionality Theory (n.d.). https://www.citelighter.com/sociology/sociology/knowledgecards/intersectionality-theory. Accessed on 12 June 2017

Intersectionality (n.d.). https://en.wikipedia.org/wiki/Intersectionality. Accessed on 12 June 2017

Sen A (2000) Development as freedom. Oxford University Press, New Delhi

Sethi M, Mittal S (2015) Gender inequities in India's urban governance. International Journal of Science and Research (IJSR) 4(8):1069–1073. ISSN (Online): 2319–7064

Appendices

Appendix 1.1: Urban Intensity Index of the Districts

Districts	Urban Intensity Index(UI)
Bankura	1.306
Bardhaman	3.107
Birbhum	1.776
Haora	2.544
Hugli	2.902
Nadia	2.538
Darjiling	1.799
Jalpaiguri	1.846
Cooch Behar	2.878
Uttar Dinajpur	1.486
Dakshin Dinajpur	1.095
Maldah	1.313
Murshidabad	2.277
North 24 Parganas	4.535
East Medinipur	1.844
West Medinipur	2.005
Kolkata	2.779
South 24 Parganas	2.523
Puruliya	1.353

Data Source: Census of India, 2011

R. Banerjee, G. Samanta, *Negotiating Terrain in Local Governance*, Local and Urban Governance, https://doi.org/10.1007/978-3-030-60663-3

Appendix 3.1: Women's Representation in the Parliament in India

Lok Sabha	Election year	Percentage of women's representation
1st	1952	4.36
2nd	1957	4.45
3rd	1962	6.27
4th	1967	5.57
5th	1971	4.05
6th	1977	3.5
7th	1980	5.29
8th	1984	7.94
9th	1989	5.48
10th	1991	7.11
11th	1996	7.36
12th	1998	7.91
13th	1999	9.02
14th	2004	8.29
15th	2009	10.86
16th	2014	11.81
17th	2019	14.39

Source: Election Commission of India

Appendix 3.2: The Comparative Representation of Women in Parliament from India and West Bengal (1952–2019)

Lok Sabha	Election year	Percentage representation of women members to total Lok Sabha members in India	Percentage representation of women members from West Bengal to total Lok Sabha members from West Bengal
1st	1952	4.36	6.45
2nd	1957	4.45	9.37
3rd	1962	6.27	6.06
4th	1967	5.57	7.69
5th	1971	4.05	5.26
6th	1977	3.5	4.76
7th	1980	5.29	4.88
8th	1984	7.94	10
9th	1989	5.48	7.32
10th	1991	7.11	7.5

(continued)

Lok Sabha	Election year	Percentage representation of women members to total Lok Sabha members in India	Percentage representation of women members from West Bengal to total Lok Sabha members from West Bengal
11th	1996	7.36	10.26
12th	1998	7.91	12.82
13th	1999	9.02	11.9
14th	2004	8.29	9.52
15th	2009	10.86	19.05
16th	2014	11.81	30.95
17th	2019	14.39	26.19

Source: www.loksabha.nic.in & www.wikipedia.org.in

Appendix 3.3: Representation of Women in Urban Governance of West Bengal

Districts	Percentage representation of women councillors to total councillors	Observed value (elected seat)	Expected value (reserved seat)	Difference between observed and expected values of representation	Percentage of observed over expected
Bankura	41.38	24	19	5	26.31
Birbhum	38.09	40	35	5	14.28
Bardhaman	38.65	80	68	12	17.64
Cooch Behar	46.25	37	26	11	42.30
Hugli	37.62	114	100	14	14
Howrah	36.49	27	24	3	12.5
Jalpaiguri	40.79	31	26	5	19.23
Maldah	40.82	20	16	4	25
Murshidabad	37.32	53	47	6	12.76
North 24 Parganas	43.27	254	194	60	30.92
Nadia	37.7	69	60	9	15
Paschim Medinipur	41.05	39	31	8	25.80
Purba Medinipur	37	37	33	4	12.12
Purulia	47.92	23	16	7	43.75
South 24 Parganas	41.18	63	50	13	26
Kolkata MC	38.89	56	48	8	16.66
Uttar Dinajpur	38.96	30	25	5	20
Darjiling	41.98	55	43	12	27.90

(continued)

Districts	Percentage representation of women councillors to total councillors	Observed value (elected seat)	Expected value (reserved seat)	Difference between observed and expected values of representation	Percentage of observed over expected
Dakshin Dinajpur	32.08	17	17	0	0
West Bengal average	39.86	56	46	10	

Source: West Bengal State Election Commission 2013–2017

Appendix 3.4: The District-Wise Patterns of the Party-Wise Representation of EWRs in Cities of West Bengal

	Party-wise percentage representation of elected women				
Districts	AITC	BJP	INC	CPIM	Others
Bankura	48.65	25	100	37.5	12.5
Birbhum	36.36	45.45	27.27	33.33	100
Bardhaman	40.91	100	30.77	36.36	16.67
Cooch Behar	71.43	0	57.14	33.33	13.64
Hugli	41.51	50	16.67	36.84	20.69
Howrah	37.93	25	66.67	39.39	0
Jalpaiguri	43.9	50	0.71	90.91	0
Maldah	56	0	50	37.5	33.33
Murshidabad	53.57	36.36	30.51	29.17	39.13
North 24 Parganas	45.09	25	33.33	40.38	33.33
Nadia	38.18		0	35.71	33.33
Paschim Medinipur	44.23	66.67	33.33	22.22	30.77
Purba Medinipur	40	0	0	0	25
Purulia	50	0	38.46	80	42.86
South 24 Parganas	45.19	25	33.33	50	50
Kolkata MC	41.23	0	20	70	12.5
Uttar Dinajpur	48.28	40	37.84	0	0
Darjiling	30.77	100	40	42.86	44.16
Dakshin Dinajpur	21.95	0	0	0	100
West Bengal average	43.96	30.97	32.42	37.66	31.99

Source: West Bengal State Election Commission 2013–2017

Appendix 4.1: Brief Profile of the Women Councillors in Darjiling

WC of wards	Age	Religion	Caste	Marital status
1	46	Hindu	General	Married
2	56	Hindu	ST	Married
5	52	Hindu	General	Married
7	38	Buddhist	ST	Married
10	39	Hindu	General	Married
13	44	Buddhist	ST	Married
14	48	Hindu	General	Married
17	32	Hindu	General	Married
20	32	Hindu	SC	Married
23	40	Hindu	General	Married
27	44	Buddhist	ST	Unmarried
29	41	Hindu	OBC	Married
30	43	Hindu	General	Widow
31	50	Buddhist	ST	Married

Source: Field Survey, 2013–2015

Appendix 4.2: Brief Profile of the Women Councillors in Balurghat

WCs of ward	Age in years	Religion	Caste	Marital status
1	60+	Hindu	General	Unmarried
3	32	Hindu	General	Married
6	63	Hindu	General	Unmarried
9	56	Hindu	General	Widow
15	44	Hindu	SC	Married
18	29+	Hindu	General	Married
21	55	Hindu	General	Widow

Source: Field Survey, 2013–2015

Appendix 4.3: Brief Profile of the Women Councillors in Raniganj

WCs of ward	Age in years	Religion	Caste	Marital status
1	55	Muslim	OBC	Widow
4	32	Hindu	ST	Married
9	45	Muslim	General	Married
12	40	Hindu	SC	Married
13	52	Hindu	General	Married
14	38	Hindu	General	Married
17	43	Hindu	OBC	Married
21	45	Hindu	ST	Married

Source: Field Survey, 2013–2015

Appendix 4.4: Brief Profile of the Women Councillors in Hugli-Chinsurah

WCs of ward	Age in years	Religion	Caste	Marital status
1	38	Hindu	SC	Married
2	42	Hindu	General	Married
9	40	Hindu	General	Married
16	38	Hindu	General	Married
17	43	Hindu	SC	Married
20	47	Hindu	General	Married
23	55	Hindu	SC	Married
25	57	Hindu	General	Married
27	50	Hindu	OBC	Married

Source: Field Survey, 2013–2015

Appendix 4.5: Profile of the Men Councillors in Darjiling

MCs of ward	Age in years	Religion	Caste	Marital status
9	46	Hindu	General	Married
12	75	Hindu	General	Married
18	44	Muslim	OBC	Unmarried
21	38	Buddhist	General	Married
25	39	Hindu	General	Married

(continued)

MCs of ward	Age in years	Religion	Caste	Marital status
26	70	Hindu	SC	Married
28	59	Buddhist	ST	Married

Source: Field Survey, 2013–2015

Appendix 4.6: Profile of the Men Councillors in Balurghat

MCs of ward	Age in years	Religion	Caste	Marital status
4	32	Hindu	General	Married
7	48	Hindu	General	Married
12	59	Hindu	OBC	Married
13	53	Hindu	SC	Married
17	73	Hindu	General	Married
23	51	Hindu	General	Married

Source: Field Survey, 2013–2015

Appendix 4.7: Profile of the Men Councillors in Raniganj

MCs of ward	Age in years	Religion	Caste	Marital status
3	46	Hindu	SC	Married
10	59	Hindu	SC	Married
15	55	Hindu	General	Unmarried
19	39	Hindu	SC	Married
20	49	Hindu	General	Married

Source: Field Survey, 2013–2015

Appendix 4.8: Profile of the Men Councillors in Hugli-Chinsurah

MCs of ward	Age in years	Religion	Caste	Marital status
3	40	Hindu	General	Married
7	45	Hindu	General	Married
10	49	Hindu	General	Married
14	62	Hindu	General	Married
18	53	Hindu	General	Married

(continued)

MCs of ward	Age in years	Religion	Caste	Marital status
22	54	Hindu	General	Married
26	48	Hindu	General	Married

Source: Field Survey, 2013–2015

Appendix 8.1: Brief Profile of the Citizens Interviewed

Cities	Categories	Brief profile of the citizens						
		Percentage of citizens under the age group of			Educational attainment of the citizens (%)			
		Less than 30 years	30 to 50 years	50 years and more	Illiterate	Nonmatric	Under graduate	Graduate
Darjiling	Men	15	50	35	–	23.5	56	20.5
	Women	12	64	24	4	36	44	16
Balurghat	Men	17	40	43	2	6	52	40
	Women	5	52	43	5	33	19	43
Raniganj	Men	24	41	35	14	55	16	15
	Women	36	52	12	24	48	22	6
Hugli-Chinsurah	Men	15	49	36	11.5	51	18	20
	Women	33	57	10	10	57	24	10

Source: Field Survey, 2013–2015

Appendix 8.2: Occupational Status of the Citizens Interviewed

Occupational status	Darjiling		Balurghat		Raniganj		Hugli-Chinsurah	
Citizens	Men (%)	Women (%)	Men (%)	Women (%)	Men (%)	Women (%)	Men (%)	Women (%)
Service	26.5	16	40	24	10	–	17	10
Private jobs	20.5	–	2	–	3	–	–	–
Cultivator	–	–	–	–	1	–	–	–
Non-earner	–	–	6	–	4	4	3	5
Housewife	–	52	–	62	–	34		66
Self-employed	17.6	16	17	9	22	3	18	
Business	35.3	16	26	5	28	3	31	
Others	–	–	9	–	32	6	31	19

Source: Field Survey, 2013–2015

Appendix 8.3: Citizens' Perception About the Leadership Quality and Behaviour of the Men Councillors

Cities	Total sample	Leadership quality (%)			Nature of behaviour (%)			
		V. good	Okay	No capability	Good	Aggressive	Busy	Without knowledge
Darjiling	30	38.09	57.14	4.76	90.48		9.52	
Balurghat	44	79.41	20.59		100			
Raniganj	44	100			97.37	2.63		
Hugli-Chinsurah	37	100			100			

Source: Field Survey, 2013–2015

Appendix 8.4: Citizens' Perception About the Work of the Men Councillors

Cities	Accessible citizens		Visit to ward				
Percentage distribution	Yes	No	Daily	Weekly	Monthly	Never	Based on problems/ sometimes
Darjiling	90.48	9.52	14.3	33.33	4.76	25.45	28.57
Balurghat	100		26.5	50	23.53		
Raniganj	100		31.5	68.42			
Hugli-Chinsurah	100		2.86	97.14			

Source: Field Survey, 2013–2015

Appendix 8.5: Perception of the Citizens About the Capability of the Men Councillors

Cities	Ability to solve problems		Given marks to councillor marks					Concerned about women empowerment	
	Yes	No	1	2	3	4	5	Yes	No
Darjiling	52.38	47.6	23.81	9.52	23.81	19.05	23.81	61.9	38.09
Balurghat	88.23	11.76			58.82	38.23	2.94	44.12	55.88
Raniganj	100		5.26	2.63	65.79	23.68	2.63	31.58	68.42
Hugli-Chinsurah	100				25.71	51.43	22.86	97.14	2.86

Source: Field Survey, 2013–2015

Index

R. Banerjee, G. Samanta, *Negotiating Terrain in Local Governance*, Local and Urban Governance, https://doi.org/10.1007/978-3-030-60663-3